MATH AND THE LEARNING DISABLED STUDENT

MATH AND THE LEARNING DISABLED STUDENT

A PRACTICAL GUIDE FOR ACCOMMODATIONS

Paul D. Nolting, Ph.D.

ACADEMIC SUCCESS PRESS, INC.
POMPANO BEACH, FLORIDA 33072

MATH AND THE LEARNING DISABLED STUDENT: A PRACTICAL GUIDE FOR ACCOMMODATIONS

BY PAUL D. NOLTING

Copyright © 1991 by Paul D. Nolting, Ph.D.

Published by Academic Success Press, Inc.

Printed in the United States of America

First Printing

First Edition

ISBN 0-940287-23-4

Table of Contents

Preface

The number of learning disabled students attending colleges and universities continues to increase. Presently there is a national movement towards increasing mathematics course requirements for certain majors or to graduate. Mathematics is already one of the most difficult college subjects for any student and usually is more difficult for learning disabled students. In fact, some colleges require students to pass up to four mathematics courses to graduate.

Many learning disabled students have received appropriate accommodations for courses that are language based such as English or social studies. Some of these accommodations are spell checkers, tape recorders and word processors. However, there is very little information on providing appropriate accommodations for learning disabled students in mathematics courses.

This book is written for counselors, mathematics instructors and administrators to provide information on appropriate accommodations for learning disabled students. Different types of learning disabilities are explained including their effects on mathematics learning. Reasons for different types of learning disabled student accommodations based on the student's disability are discussed. Testing accommodations are suggested in accordance with the student's learning disability. The importance of these accommodations is to make sure that mathematics knowledge is tested instead of the student's disability.

Practical suggestions to help learning disabled students become more successful in mathematics are discussed in case histories. These five learning disabled student case histories range from simple to complex. A learning disability profile is presented for each student along with the accommodations used. The accommodations used in these cases can be generalized to other learning disabled students.

Helping learning disabled students become successful in mathematics is the responsibility of administrators, counselors, and mathematics instructors. This group needs to work together to provide adequate financial resources, counseling for disabled students, and appropriate accommodations. More learning disabled students can succeed in mathematics by following the suggestions in *Math and the Learning Disabled Student.*

Chapter 1

Learning Disability Definitions

1

Community colleges and universities are experiencing a substantial increase in learning disabled students. Improved special education programs in elementary schools, middle schools, and high schools have helped learning disabled students graduate. White, Alley, Deshler, Schumaker, Warner, and Clark (1982) estimate that about 67 percent of high school students with learning disabilities plan to attend college. In the past, learning disabled students had special education classes with direct teacher instruction. Now, without separate special education classes, many learning disabled students are having major difficulties adapting to college-level courses. In addition, mature students are making up a higher percentage of college students. Many of these mature students had previous learning problems in school, but existing programs could not diagnose them as learning disabled. Jarrow (1987), Director of the Association on Handicapped Student Service Programs in Postsecondary Education, has stated that learning disabled students are "the single largest contingent of students with disabilities being served on American campuses" (p.46). Since mathematics is considered education's *critical filter,* learning disabled students need additional learning skills and accommodations in mathematics to reach their education potential.

Hammill (1990) used twenty-eight recent textbooks that focused on learning disabilities to review current definitions of learning disabilities. The textbook sources referred to eleven different learning disability definitions that are being or have been used. In describing learning disabilities there is a clear distinction between a conceptual definition and an operational definition of learning disabilities. Conceptual definitions focus on the theoretical aspects of a learning disability and should clearly state the exact aspects of a learning disability. Operational definitions are based on conceptual definitions and can be used to qualify students as learning disabled. Operational definitions usually have a statistical requirement that the student's test scores must meet to qualify the student as learning disabled. Operational definitions can vary widely from college to college and state to state. Some of the conceptual definitions are described below for the reader to obtain a better understanding of learning disabilities.

The term learning disability (LD) is usually used to describe a broad range of neurological dysfunctions. Since learning disabilities are invisible this term is often misunderstood. The Federal Register defines a LD as:

> A disorder in one or more of the basic psychological processes involved in understanding and use of language, spoken or written, which may manifest itself in an imperfect ability to listen, think, speak, read, write, spell, or to do mathematical calculations. The term includes such conditions as perceptual handicaps, brain injury, minimal brain dysfunction, dyslexia, and developmental aphasia. The term does not include individuals who have learning problems which are primarily the results of visual, learning or motor handicaps, or mental retardation, or environmental, cultural, or economic disadvantage.

A learning disability, according to the Title V regulations in the California community college system is:

> ... a persistent condition of presumed neurological dysfunction which may also exist with other disabling conditions. This dysfunction continues despite instruction in standard classroom situations. Learning disabled adults, a heterogeneous group, have these common attributes:
>
> a) average to above average intellectual ability
> b) severe processing deficits
> c) severe aptitude-achievement discrepancy(ies)
> d) measured achievement in an instructional or employment setting
> e) measured appropriate adaptive behavior in an instructional or employment setting

A specific learning disability, according to rule 6H-1.041 for the Florida community college system, is:

> A disorder in one or more of the basic psychological or neurological processes involved in understanding or in using spoken or written language. Disorders may be manifested in listening, thinking, reading, writing, spelling, or performing arithmetic calculations. Examples include dyslexia, dysgraphia, dyscalculia, and other specific learning disabilities in the basic psychological or neurological process. Such disorders do not include learning problems which are primarily due to visual, hearing, or motor handicaps, to mental retardation, to emotional disturbance, or environmental deprivation.

Hammill (1990) states that the 1977 U.S. Office of Education learning disability definition and the National Joint Committee on Learning Disabilities (NJCLD) definition are the two most often used conceptual definitions. The 1977 U.S. Office of Education definition is popular because of its legislative support. The NJCLD definition is popular because of its support from several different national organizations who represent learning disabled individuals. According to Hammill (1990) the NJCLD is the best definition of a learning disability.

The 1977 U.S. Office of Education definition is:

> The term "specific learning disability" means a disorder in one or more of the basic psychological processes involved in understanding or in using language, spoken or written, which may manifest itself in an imperfect ability to listen, speak, read, write, spell, or to do mathematical calculations. The term includes such conditions as perceptual handicaps, brain injury, minimal brain dysfunction, dyslexia, and developmental aphasia. The term does not include children who have learning disabilities which are primarily the results of visual, hearing, or motor handicaps, or mental retardation, or emotional disturbance, or environmental, cultural, or economic disadvantage. (USOE, 1977, p. 65083)

One of the most widely accepted definitions of learning disabilities was developed by the National Joint Committee on Learning Disabilities (1988):

> Learning disabilities is a general term that refers to a heterogeneous group of disorders manifested by significant difficulties in the acquisition and use of listening, speaking, reading, writing, reasoning, or mathematical abilities. These disorders are intrinsic to the individual, presumed to be due to central nervous system dysfunction, and may occur concomitantly with other handicapping conditions (for example, sensory impairment, mental retardation, serious emotional disturbance) or with extrinsic influences (such as cultural differences, insufficient or inappropriate instructions), they are not the results of those conditions or influences.

A general description of someone with a learning disability would be a person with average to above average intelligence who has difficulty in one or more of the basic neurological functions such as perception (auditory, visual or spacial). This disorder may impair reading (dyslexia), writing (dysgraphia), mathematical calculations (dyscalculia), thinking, and spelling. This disorder does not include learning problems primarily due to physical disabilities, emotional disturbance, or lack of previous opportunity for learning, even though some learning disabled students may become physically disabled or have social or emotional problems. In some states the definition for a mathematics learning disability would include a severe difference between mathematics or general aptitude scores and mathematics achievement scores. Learning disabilities can not be "cured". But in many cases learning disabilities can be circumvented through learning and testing accommodations.

Students who may have a mathematics learning disability might experience these symptoms:

a) Difficulty doing the actual calculations
b) Difficulty learning a series of mathematics steps to solve a problem
c) Inability to apply mathematics concepts to word problems
d) Difficulty solving oral problems

However, these same symptoms may be the result of other types of learning disabilities such as visual processing speed or auditory processing disorders.

Chapter 2

Determining the Existence of a Learning Disability

2

In order to fully understand the appropriate accommodations for students with learning disabilities, it is appropriate to review the technical aspects of various procedures used to determine if a learning disability exists. According to An Instructional *Guide to the Woodcock Johnson Psycho-Educational Battery* (Mather, 1991), after a student meets the conceptual definition of a learning disability there are three different procedures for documenting a possible learning disability. The first indicator of a learning disability is a significant aptitude-achievement discrepancy. The second indicator is a significant intracognitive discrepancy. The third indicator of a learning disability is a significant intra-achievement discrepancy between academic levels.

Other professionals identify a learning disability by comparing the student's intelligence score to achievement scores or the different cognitive cluster scores. There does not appear to be one best way to identify a learning disability. For your own information, consult your state department of Vocational Rehabilitation to find out how they determine if a learning disability exists.

One procedure to identify a mathematics learning disability is to look for a significant discrepancy between the student's mathematics aptitude and the student's mathematics achievement. This is the traditional procedure. The mathematics aptitude score is usually measured by certain cognitive subtests of the *Woodcock Johnson Psycho-Educational Battery - Revised* (WJ-R) (1989). These tests include Visual Matching, Analysis-Synthesis, Oral Vocabulary, and Concept Formation. The Visual Matching test consists of identifying and circling identical numbers among a row of different numbers. The Visual Matching test measures the student's speed in recognizing two of the same numbers and motor speed required to circle the numbers. The Analysis-Synthesis test measures the student's ability to figure out the missing parts of a mathematics related logic puzzle, which is part of fluid reasoning. The Oral Vocabulary test consists of synonym and antonym tests. It measures a student's comprehension and is the best estimate of overall intelligence. The Concept Formation test is a measurement of reasoning ability which has the student recognize, use, and explain rules to logic puzzles. The Concept Formation test is a part of Fluid Reasoning cluster tests. In general, mathematics aptitude measures number processing and verbal and nonverbal abstract reasoning. Mathematics aptitude score is used to predict the student's expected mathematics achievement level.

Mathematical achievement is mainly measured by the Broad Mathematics cluster of the *WJ-R* or other accepted mathematics achievement tests. The Broad Mathematics cluster consists of the Calculation subtest and the Applied Problems subtest. Applied Problems subtest consists of story (word) problems.

To document a significant aptitude and achievement discrepancy, compare the standard aptitude score to the standard Broad Mathematics achievement cluster score on the *WJ-R*. The mathematics aptitude score must be higher than the mathematics achievement score. The standard score is the most often-used comparison score that makes scores from different tests comparable. The mean or average for standard scores is 100 with a standard deviation of 15 points. The differences between the aptitude and achievement scores are measured in standard deviations. For example, if two scores are two standard deviations apart, that would mean a 30 point difference between the aptitude and achievement scores.

Morris and Leuenberger (1990) have a three-step process operational definition of a learning disability. The three-step criterion are:

1. Average or above average intelligence score (Full Scale IQ of 85 or greater)
2. 1.3 standard deviation difference between ability and achievement score
3. No evidence of sensory, mental, emotional, cultural, or environmental problems

Salvia, Gajar, Gajria, and Salvia (1988) use a 40 point percentile difference between mathematics achievement and ability to classify students as learning disabled.

In general there is no set standard score difference or standard deviation difference required between two scores which determines a learning disability. Colleges, universities or state agencies usually select a difference between 15 to 30 standard score points or between one and two standard deviations difference between aptitude and achievement scores as part of the operational definition of a learning disability. Hessler (1984) suggests that a measure of general intelligence such as by the *Wechsler Adult Intelligence Scale - Revised* (1981) is a better predictor of academic achievement than an aptitude test. Mather (1991) suggests that the aptitude/achievement comparison is a controversial procedure to document a learning disability even though it is widely used.

Intracognitive cluster discrepancies among the different cognitive subtests may also identify a learning disability resulting in mathematics learning problems. Batchelor, Gray & Dean (1990) have documented that both verbal and nonverbal neuropsychological functions are related to arithmetic performance. Requisites for arithmetic learning appear to be attention, concentration, verbal-auditory discrimination, and visual tracking. There was a significant correlation between continuous visual stimulus and mathematics achievement. Visual stimulus was accounted for by verbal facility, nonverbal intermediate memory, and verbal abstract reasoning. Their conclusion indicates that deficits in the following areas may cause arithmetic learning problems: attention, concentration, verbal-auditory discrimination, visual-spacial processing, psychomotor functioning, nonverbal concept formation, and perceptual functioning. Future researchers should continue to determine the relationship of visual and auditory processing to mathematics learning.

Some of these neuropsychological functions that were measured in the Batchelor article are part of the intracognitive clusters discussed below. To document an intracognitive discrepancy, compare the standard scores of the following *WJ-R* cluster scores: Long-Term Retrieval, Short-Term Memory, Processing Speed, Auditory Processing, Visual Processing, Comprehension-Knowledge, and Fluid Reasoning. To determine the exact difference between each score used to determine a learning disability, contact your Office of Disabled Students or testing center.

Intra-achievement cluster discrepancies among different achievement cluster scores may identify the existence of a learning disability. The achievement areas are: Broad Reading, Broad Math, Broad Written Language, and Broad Knowledge. As stated before, contact your Office of Disabled Students or testing center to find out the exact difference needed among the standard test scores to determine if a learning disability exists.

Comparing a student's intelligence score to *WJ-R* cognitive cluster scores or achievement scores is probably the most often used process to identify a learning disability. The student's intelligence is usually measured by the Wechsler Adult Intelligence Scale - Revised (1981). Mathematics achievement is usually measured by the *WJ-R* Broad Math score. Other intelligence tests and mathematics achievement test scores can be used for comparison. The student's IQ score can also be compared to the student's *WJ-R* cognitive test cluster scores such as Visual Processing or Short-Term Memory.

Contact the Office of Disabled Students or testing center at your school to find out the difference required between the IQ standard score and cognitive cluster standard score or mathematics achievement standard score to determine if a learning disability exists. Remember, even though the difference between a student's test scores meets the college/university's statistical requirement, the student still must meet the conceptual learning disability definition.

Chapter 3

Reasons for Mathematics Learning Problems

3

Learning disabled students who have mathematics learning difficulties may have a mathematics learning disability. However, in my experience, most of these students have major information processing disorders. These processing disorders block their ability to obtain valuable information to learn mathematics and/or to demonstrate their mathematics knowledge on tests. Most of these learning disabled students did not have a severe difference between their general or mathematics aptitude/IQ and mathematics achievement scores. When the effects of their learning disability were diminished by appropriate learning and testing accommodations most of their mathematics grades improved.

Students with a mathematics learning disability (a significant discrepancy between mathematics aptitude/IQ score and mathematics achievement score) may not understand the reasons for their mathematics learning problems. These students' mathematics aptitude tests score or IQ scores are significantly higher than their mathematics achievement score. However, these test scores or the subtest scores usually do not reveal all the reasons for mathematics learning problems.

For example: the mathematics aptitude score does not measure short-term memory, auditory processing or visual processing skills which are related to mathematics learning. Students with these types of processing problems could have good abstract reasoning but have difficulty understanding mathematics lecture information.

To understand the learning problems of a mathematics learning disabled student, each intracognitive cluster score needs to be reviewed. Potential intracognitive cluster problem areas can be divided into processing, fluid reasoning, and long-term retrieval. Learning accommodations and testing accommodations for mathematics learning disabled students must also be based on processing, fluid reasoning, and long-term retrieval disorders.

Processing Disorders

As different processing disorders are discussed, pay special attention to the processing disorders that are evident with your learning disabled students. If you do not know the type of processing disorders affecting your students refer to the students' test records, or refer them for an assessment. If you are unable to find out the type of processing disorders of your students, then read these descriptions of different processing disorders. Pick the processing disorder that is similar to your students' learning problem. **Remember: only by having your students go through an**

assessment program can you really know the type and extent of their processing disorder(s).

Learning disabled students who have visual processing speed disorders will have difficulty learning mathematics. Batchelor et al. (1990) reported that visual tracking, which is similar to visual processing speed, is a very important factor in mathematics achievement. In the *WJ-R* the Visual Processing Speed cluster consists of two tests. These two tests measure the student's ability to rapidly scan and identify visual material. Visual processing speed is the speed of working with understood mathematics symbols and numbers. In other words, this is the speed at which students can copy down recognizable numbers and symbols.

A student's visual processing speed will affect how fast he/she can copy notes from the board and take a mathematics test. Kirby and Becker (1988) discovered that students having mathematics learning problems were slow in doing their mathematics problems. These students had slow visual processing speed. Many of these students performed accurate mathematics computations but could not do them quickly.

Learning disabled students who have visual processing disorders will have trouble quickly identifying symbols. Rourke and Strang (1983) and Strang and Rourke (1985) identified neuropsychological distinct subtypes of arithmetic dysfunction. One type of arithmetic dysfunction was with students who had normal achievement in areas other than arithmetic. These students had normal verbal and auditory-perceptual skills, but demonstrated deficiencies in both visual-spacial and tactile-perceptual skills. These students also had difficulties with concept formation and problem solving.

According to the *WJ-R*, the Visual Processing cluster consists of two tests. These two tests measure the student's ability to recognize masked pictures from a group of several pictures. The Visual Processing test measures a student's ability to recognize and think in visual designs. In other words, it measures a student's ability to recognize and remember, in sequence, complex mathematics symbols and numbers that may not be known. For example, a student's ability to recognize and remember this polynomial:

$$4x^2 + 2x + 1$$

Learning disabled students may have difficulty telling the difference between 2 as part of a factor or an exponent, and between + as a plus sign, or x as a variable. Remembering the correct parts of the polynomial in order can be a problem. For instance, remembering that the 4 rather than the 2 is the coefficient of the squared term. This will cause major problems when the student copies material from the board. It also will cause problems reading tests and the mathematics textbook. Mistakes can occur when students miscopy notes, and misread the textbook or test questions.

Learning disabled students with auditory memory difficulties such as short-term memory difficulties will have difficulty learning mathematics. The *WJ-R* cognitive cluster for Short-Term Memory consists of two tests. These two tests measure a student's ability to recall sentences or a list of unrelated words. Short-term memory difficulties can be manifested in remembering numbers, symbols or words in their correct order. This problem becomes apparent when listening to a mathematics instructor explain the steps of working a problem. For example: students may forget the mathematics problem steps before writing them in their notes. Also, students could recall and write down the mathematics problem steps in the wrong order. Either mistake will cause difficulty in understanding the mathematics problem and using class notes as a homework guide.

Students with low performance on the Short-Term Memory cluster will have difficulty reading the mathematics textbook (Badian, 1988). These students have difficultly retaining mathematics concepts long enough in short-term memory to understand the concept. If a series of related mathematics concepts are discussed in the mathematics textbook these students will have difficulty remembering one concept long enough to apply it to the next concept.

Students with poor short-term memory will most likely have difficulty with mathematics word (story) problems. These students will spend more time re-reading the word problems to understand the question. After understanding the question they will spend additional time setting up the equation to solve the problem. In general, students with poor short-term memory have difficulty in most of their reading areas. According to Blalock (1980), about one-third of young adults with learning disabilities have some type of auditory memory problem.

Students with auditory processing difficulties have problems telling the difference between certain sounds of words. Rourke and Strang (1983) and Strang and Rourke (1985) discovered another type of neuropsychological arithmetic dysfunction was with students with normal visual-spacial and tactile-perceptual skills. However, these students had deficits in verbal and auditory-perceptual skills which are similar to auditory processing problems. The measurement of auditory perceptual skills is similar to the tests used to measure auditory processing.

The *WJ-R* uses two subtests as part of the auditory processing cluster. These two tests measure the student's ability to recognize words with missing sounds and the ability to put together a group of sounds into a word. Auditory processing is a measurement of a student's ability to put together different sound patterns into words and evaluate the difference between auditory patterns.

There is a definite relationship between auditory processing and reading achievement. Wanger (1986) reviewed about 10 years' of research on phonological processing and reading disabilities. He discovered that one of the causes of poor reading achievement is poor auditory processing. Kochnower, Richardson, and DiBenedetto (1983) discovered that children with learning disabilities could not phonetically read words compared to students without learning disabilities. That means that some learning disabled students will have poor reading levels due to auditory processing deficits. This is just another reason why learning disabled students have difficulty understanding the mathematics textbook. The learning disabled student has to focus more on the lecture to obtain the knowledge to understand the mathematics concepts.

Adult learning disabled students may continue to have difficulty understanding tasks or concepts that require good auditory processing skills. Meyers (1987) explains that not only can reading efficiency be affected by an auditory processing deficit, the student's ability to understand lectures and oral directions is dramatically affected as well. In fact, the more severe the auditory processing deficits, the more the student can misunderstand the lecture material. The student will "miss" some of the words in a lecture or replace words with incorrect words. When this occurs, the student will have difficulty understanding the instructor and writing his/her notes. Meyers continues to explain that these students will have major difficulty understanding lectures in distracting situations. Either auditory processing problems during reading of the mathematics textbook or listening to lectures can cause the student to misunderstand mathematics concepts.

In general, processing disorders hinder students in receiving lecture information and taking tests. The lecture information could be learned in the incorrect order or totally forgotten before the information is recorded as notes. Processing difficulties, especially when taking a mathematics course, can cause problems learning a sequence of mathematics concepts.

Students who have above average grades in all their courses but mathematics presents a special concern. These students usually make "D"s or "F"s in their mathematics courses but make "A"s or "B"s in their other courses. Many of these students may have a learning disability along with poor mathematics study skills and test anxiety. The fact that some of these students are considered gifted in their specialty areas such as humanities, art, English and athletics further confuses the problem.

Waldron and Saphire (1989) compared the intellectual patterns of gifted learning disabled students to gifted non-learning disabled students. The gifted learning disabled students performed significantly poorer in perceptual areas such as visual and auditory discrimination, visual and auditory sequencing, short-term auditory memory and visual-spacial skills compared to the non-learning disabled gifted students. No difference existed between the groups in visual memory skills or listening comprehension. The gifted learning disabled students also had comparative weakness in mathematics, reading, and spelling. Waldron and Saphire (1989) concluded that many of the gifted learning disabled students' academic difficulties may have been caused by perceptual problems.

Daniels (1983) compared the strengths of gifted learning disabled students using the *WISC-R* and discovered that their highest scores were on the Similarities test. The Similarities test measures reasoning and conceptual thinking ability. Their lowest scores were on the Arithmetic and Digit Span tests. The Low Arithmetic test scores may indicate poor mathematics skills or reasoning and an inability to concentrate or listen. The Low Digit Span test scores may indicate poor short-term memory, poor attention or lack of concentration. In summary, gifted learning disabled students tend to have problems in rote auditory memory and auditory discrimination between similar sounding words and letters.

Waldron and Saphire (1990) continued their research with gifted learning disabled students by comparing their learning characteristics to gifted non-learning disabled students. The gifted learning disabled students were strongest in verbal conceptualization, reasoning in processing information, and thinking skills. They were significantly weaker in rote recall of verbally presented information. Their ability to mask short-term memory problems is probably due to their verbal conceptualizing skills. This research also supports the finding in the study by Daniels (1983). In addition Bauer (1982) states that gifted learning disabled students may have alterations in their basic cognitive processes. They may be having problems in one or more of the four stages of processing: encoding, manipulation, response selection, and response execution. He also suggested that poor short-term memory or long-term memory may affect learning patterns in these students.

The research on gifted learning disabled students can give us insight into the learning problems of students who make good grades in everything but mathematics. The processing and memory problems exhibited by these students are emphasized in the learning of mathematics. Gifted or non gifted students may "live off" certain aspects of their high IQ (learning ability) to achieve a certain level of mathematics competency. However, these students may eventually take a mathematics course where their IQ (learning ability) will not have enough "power" to compensate for their other learning problems. Then these students start failing their mathematics courses. Instructors usually start questioning the students' effort and motivation in learning mathematics. The students then start questioning their own ability which can start the self-doubt syndrome. To help these students we must understand how processing deficits and memory problems affect each student's mathematics learning.

Fluid Reasoning

Another type of learning disability is manifested in problems with fluid reasoning or thinking. These problems usually occur when abstract reasoning is required to apply some type of mathematics concept from its property or principle to its application. The *WJ-R* Fluid Reasoning cluster is measured by the Analysis-Synthesis test and the Concept Formation tests. The Analysis-Synthesis test measures the student's ability to figure out the missing parts of a mathematics-related logic puzzle. The Concept Formation test measures the student's ability to recognize, use, and explain rules to logic puzzles. In short, the Fluid Reasoning cluster measures nonverbal abstract reasoning and problem solving skills. For example, students may have difficulty understanding and applying a formula to a new homework problem. Students may remember the concept but forget how to use it in a problem. In general, students may have difficulty understanding abstract mathematics formulas and generalizing the formula's use to their homework or test problems.

Hessler (1984) indicates that the performance on the reasoning score (fluid reasoning) is very closely related to mathematics achievement. Students who score high on the Fluid Reasoning cluster have good mathematics achievement scores. Students who score low on the Fluid Reasoning cluster will, in most cases, do poorly on mathematics achievement. In addition, students who score low in the Fluid Reasoning cluster frequently will have difficulty solving problems that require conceptualization and abstract mental processing. Some of the courses that students may have difficulty in are biology, chemistry, geometry, and physics.

Bley and Thornton (1981) suggest that students with learning disabilities in fluid reasoning may have more trouble with mathematics than any other area. Learning disabled students with reasoning problems often have difficulty verbalizing what has been learned, relating the concepts with symbolic language, and auditorially or visually understanding the instructor's explanation.

Fluid reasoning is an excellent predictor of student success in algebra. Teaching students how to improve their fluid reasoning is difficult. Fluid reasoning, in adult students, may also be related to reading comprehension (Mather, 1991).

Long-Term Retrieval

Some learning disabled students have difficulty with long-term memory. Long-term memory is a component of mathematics learning that acts as an information source for mathematics formulas, mathematics vocabulary, the recognition of different problem types, algorithms, and heuristics. Deficits in long-term memory result in poor arithmetic achievement (Dinnel, Glover, Ronning, 1984).

The *WJ-R* measures long-term memory by using the Long-Term Memory Retrieval cluster. The Long-Term Memory Retrieval cluster measures a student's efficiency in storing and recalling information over a period of time. The Long-Term Memory Retrieval cluster consists of two pre- and post-tests. The first pre-test measures the student's ability to remember and recall novel names through auditory-visual association. The second pre-test measures visual-auditory association learning by having the student remember nonsense symbols in the form of sentences. The two post-tests measure the student's ability after several days to recall novel names or symbols.

For example, students may be inconsistent when learning new facts or concepts. Students might be able to learn how to work fractions one day and a week later have difficulty recalling how to work fractions. However, when they are taught fractions again, it is easier for them to learn the steps.

The same memory loss can happen working multiplication tables. Students may forget the multiplication tables, but demonstrate the concept of multiplication. Students may have poor achievement in mathematics calculations, but may have average or above average mathematics reasoning ability. Usually the result of a poor long-term memory is that one time the student understood how to work a problem but forgot the concept or mechanical calculation needed to solve the problem.

Mather (1991) indicates that students with long-term retrieval problems have difficulty with paired-association tasks such as remembering the letters of the alphabet or their multiplication tables. This also suggests future difficultly in reading and mathematics.

Stages of Memory

Nolting (1988) suggested how learning can be affected by the stages of memory. For an overall understanding of how learning disabilities affect sensory input, learning, memory, and testing, a review of the stages of memory is helpful. The stages of memory are an expansion of Atkinson and Shiffrin's(1968) model. The stages of memory are sensory input, sensory register, short-term memory, long-term memory and memory output (Figure 1). These stages of memory learning processes are presented as a simplified theoretical model of learning. A student's learning disability can affect the processing of information through any of these memory stages.

Sensory input is the form in which the information is received by the student. A student's sensory input can be in the form of hearing, seeing, feeling, smelling or touching the information. In a mathematics course the student's sensory input of information is mostly by hearing a lecture and is supplemented by seeing the information written on the board.

The sensory register holds the exact image, sound, smell, touch, or feeling of the sensory experience until it can be processed. This raw sensory information is encoded so it can be used in the next stages of memory. The sensory register holds visual data for about one second and auditory data for about four seconds. If this information is not processed immediately it fades out and is forgotten (Bourne and Ekstrand, 1985).

The learning disabled student's sensory register may be dysfunctional in regard to the speed it encodes information or the accuracy with which it encodes the information. Learning disabled students with visual processing speed problems and auditory processing deficiencies may not receive all the information or the correct information through the sensory register. This information may have to be reprocessed through the sensory register several times to obtain the correct information. If the information obtained through the sensory register is incorrect, it severely affects the remaining memory stages, leading to incorrect learning. For example: continuously reading a multiplication sign (x) as an addition sign (+) while doing arithmetic problems. Information processed slowly or incorrectly through the sensory register can cause problems in note-taking, reading a mathematics textbook and timed tests.

Information processed through the sensory register is transferred into short-term memory. Short-term memory is a temporary storage of information that enables the recall or recognition of that information. Short-term memory fades out in about 20 to 30 seconds and sometimes lasts only five seconds. The capacity of short-term memory is seven bits of information plus or minus two information bits. Some examples of short-term memory are remembering telephone numbers and people's names. Information has to be held in short-term memory long enough to understand it. If information is not rehearsed it will quickly fade from short-term memory.

Learning disabled students who do poorly on the Short-Term Memory cluster have difficulty retaining auditory information long enough for it to be processed. During a lecture these students may not be able to retain the information long enough to write it down in note form. Other students cannot hold the information long enough to understand the mathematical concept. Since learning mathematics is a step-by-step process, missing only one step due to a short-term memory problem is disastrous.

The *WJ-R's* Visual Processing test in part measures a student's short-term memory ability to retain visual information. Learning disabled students with visual processing problems will have difficulty remembering complex visual information long enough to transfer it into notes. These students may have major problems trying to recognize and retain complex mathematics problems on a test. Any short-term memory problem can disrupt the learning and testing process.

The next stage of memory is called long-term memory. Long-term memory consists of information processed through short-term memory that has been rehearsed. Long-term memory is the student's total knowledge that goes beyond basic facts and ideas. Long-term memory is the organization of short-term memories into meaningful information, thinking about them, and comprehending the meaning. One of the main problems that students have in learning is transferring short-term memory into useable long-term memory.

According to the *WJ-R,* students with long-term retrieval problems have difficulty storing and retrieving information over a long period of time. Learning disabled students with this problem may know mathematics properties one day and several weeks later not remember them. This problem also occurs with students who cannot retain their multiplication tables. These student might do fine on mathematics quizzes but fail their major mathematics tests. Since learning mathematics is accumulative, these student usually do worse on each successive mathematics test. However, in some of their other subjects which do not have accumulative tests, these same students might be making "A"s and "B"s.

Memory output is the last of the stages of memory. Memory output is the student's ability to correctly process a mathematics test, recall the problem steps and write the answers. Since most mathematics tests are timed, the mathematics test will measure mathematics knowledge and the severity of the student's learning disability.

For example: learning disabled students with visual processing speed or visual

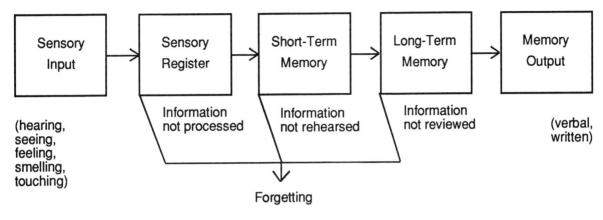

Figure 1
Stages of Memory

processing problems will take longer just to understand the test questions compared to other students. In some cases these students will misinterpret the test question and/ or their own problem solving steps. When this occurs, the mathematics test is measuring their inability to visually understand and process the test. Any of these problems that a learning disabled student has usually negatively affects his test results.

Reviewing stages of memory from a theoretical base can aid in understanding how learning disabled students have difficulty learning information and recalling that information for mathematics tests. This model can also be used as a basis for understanding the appropriate learning and testing accommodations required for learning disabled students to have equivalent learning oportunities as compared to non-learning disabled students.

Linear Learning

Learning disabled students taking mathematics classes usually experience more difficulty learning mathematics than other subjects. One major reason is that mathematics requires linear learning. This means that material learned one day is used the next day and the next month. Learning disabled students with short-term and/or long-term memory problems can have difficulty learning mathematics. These same students in many nonlinear courses, where material learned one day can be forgotten after the test, will not have as much difficulty learning the new material.

For example: in some non-linear courses--such as the social sciences--misunderstanding the first part of a lecture doesn't mean students won't understand the rest of the class material. However, in a mathematics class, if students misunderstand the first concept taught, they are probably lost for the remainder of the class. In a non-linear class students can probably read the textbook to obtain the missed information. Due to the complexity of the mathematics textbook, students probably won't be able to obtain missed information from the textbook. These are several reasons why learning disabled students may have problems learning mathematics compared to learning non-linear subjects.

These learning problems are not characteristic of all learning disabled students. For example:I worked with a learning disabled woman whose major learning problem was poor visual processing speed. She also had a short-term memory problem. She had difficulty following directions and had high test anxiety. Her most positive learning aspect, like many other learning disabled students, was persistence and an achievement orientation for success. A student's learning disability does not just affect his/her education. It manifests itself in many signs and symptoms across the student's life.

This discussion on the reasons that learning disabled students have difficulty learning mathematics has focused on specific learning disabilities. Some learning disabled students have a combination of two or more learning disability disorders. Learning disabled students can also have additional learning problems that are a result of their learning disability. In general, learning disabled students may have problems in some of the following areas:

  a) Understanding humor
  b) Following oral directions
  c) Math or general study skills
  d) Time management
  e) Developing positive study attitudes
  f) Developing a positive self-image
  g) Concentrating with on-going distractions
  h) Recalling everyday words
  i) Socially talking to students
  j) Having high test anxiety

These additional learning problems also affect the student's ability to learn mathematics.

Chapter 4

Mathematics Learning Accommodations

4

Learning accommodations allow learning disabled students the same access to course material as non-disabled students. Course materials can be in the form of lectures, labs, field trips, library resources, and books. Learning accommodations may be the same or different for various types of learning disabled students.

Visual Processing Speed/Visual Processing

In my opinion, most learning disabled students with mathematics learning problems have visual processing speed and/or visual processing disorders. In fact, with my learning disabled students, visual processing speed disorders are the most common mathematics learning problem. Dr. Hessler (1984), author of *Use and Interpretation of the Woodcock-Johnson Psycho-Educational Battery,* also suggests that visual processing/speed is more highly related to mathematics achievement than other measures of perception. Many of these learning disabled students are in the first to tenth percentile on Visual Processing Speed or Visual Processing tests. This means that 90 to 99 percent of the other students visually process material faster or better than the learning disabled student. For these students, taking mathematics notes is a major problem. They usually take very few notes or try to write everything down. They can have number reversals or transpose parts of an equation while writing notes. They may also have difficulty keeping problem steps properly aligned in the correct columns.

Learning disabled students who write down a few notes don't concentrate as much on the lecture and have little information for solving mathematics homework problems. Other learning disabled students spend all their concentration on note-taking instead of listening to the instructor. These students end up writing another "mathematics textbook" instead of understanding mathematics concepts by listening. **Remember: in most cases the mathematics instructor can better explain mathematics concepts than the textbook. Both of these note-taking strategies leave the student with mathematics learning gaps.**

Three lecture accommodations for learning disabled students with visual processing speed or visual processing problems are note-takers, tape recorders, and handouts. Learning disabled students may need only one or all of these learning accommodations. Learning disabled students may only need these accommodations in a mathematics class and not in their other subjects. Learning accommodations depend on the learning disabled student's special needs.

When selecting note-takers, try to get a student that is in the learning disabled student's mathematics class. Have a potential note-taker give a sample of his/her notes to the learning disabled student to see if the note-taking style meets the learning disabled student's needs. The note-taker can use NCR paper, carbon paper or make a copy of his/her notes for the student's use. The learning disabled student still must take some notes to maintain concentration and to be used as future guidelines. Have the learning disabled student leave four or five lines between lecture information gaps. The learning disabled student can fill these gaps by recopying the notes from the note-taker. However, some learning disabled students may find note-taking easier by turning their notebooks sideways. This procedure creates vertical columns for numbers or variables. Each number or variable can be placed into one column for easy tracking. Another idea is to use enlarged graph paper so that the student can use both the vertical and horizonal lines for tracking.

Learning disabled students must copy important information from the note-taker's notes to their notes on the same day. This procedure will help them understand and memorize notes more quickly. The most important factor is to have good listening and note-taking techniques like those discussed in *Winning At Math* (1991). Note-takers should be taught these listening and note-taking skills as well!

Learning disabled students might want to use a tape recorder to review the lecture. The tape recorder must have a tape counter to be useful. During a lecture the instructor will start working a problem on the board. If this problem becomes difficult, have the student write down the tape counter number and concentrate on the instructor's explanation. Once the instructor finishes the problem, the student can write down the tape number. As soon as possible have the student play back that part of the tape and fill in his/her notes. If the student still doesn't understand the problem, have him/her go to the instructor or tutor for help. Remember: learning disabled students must re-work their notes as soon as possible to get the most benefit from them. Use the reworking-your-notes techniques discussed in *Winning At Math* (1991).

Handouts can improve mathematics learning. Handouts can decrease the amount of note-taking necessary, list important information, and improve students' concentration during lectures. Students can ask instructors for future handouts or for additional handouts on current class materials. Students can also make their own handouts by copying and enlarging mathematics pages that have rules and concepts. Have the students use these copies as a guide in the next mathematics class and even take notes on them.

Learning disabled students with visual processing speed and visual processing disorders will have difficulty reading their mathematics textbooks. Engelmann, Carnine, and Steely (1991) suggest that mathematics textbooks seem to be linked to poor mathematics performance and are difficult for learning disabled and at-risk students to understand. This problem is further compounded by the fact that most students use the same reading procedures for their mathematics textbook as they do for other textbooks. They do not realize that reading a mathematics textbook requires a different approach. Reading a mathematics textbook requires rapid recognition and an understanding of words intermixed with symbols. Students with visual processing problems become easily frustrated with mathematics textbooks due to the various symbols and abrupt structure changes. Students with visual processing speed problems who may already read extremely slow, also become frustrated because of the extended time required to read the mathematics textbook. **No wonder most disabled and non-disabled students do not read their mathematics textbooks!**

Another problem in reading a mathematics textbook is that textbook pages are so cluttered it is like going through a visual mine field, and many students don't survive. Many learning disabled students must be taught how to read a mathematics textbook or have someone explain the textbook information to them. However, mathematics instructors lack the skills required to teach students reading techniques. Learning disabled students must develop reading techniques to understand their mathematics textbook. To understand mathematics concepts, learning disabled students must also learn the vocabulary and symbols that are specific to mathematics.

A good reading technique can improve understanding of the mathematics textbook. Students can use the eight-step mathematics reading technique discussed in chapter six in *Winning At Math* (1991) to improve understanding and retention of the mathematics material. In some cases, an enlarged photocopy of a page with formulas may be helpful. Using different colored pens or magic markers to highlight parts of equations that have different meanings can also be helpful. If these or other reading techniques do not work, then have the textbook put on audio tape. However, a mathematics textbook on audio tape requires a lot of concentrated listening and can sometimes be ambiguous due to the presence of formulas and graphs. To improve students' learning of the mathematics textbook, have them listen to the audio tape as they read the textbook.

In many cases, a tutor will be required for students with visual processing speed and visual processing problems. Even with the already-mentioned learning accommodations, a tutor may have to go over some of the more difficult mathematics concepts or problems such as graphing equations. The tutor should use different colored pens or highlighters to indicate exponents, x and y axes or sign changes. The tutor should also have some training on the characteristics of learning disabled students. The tutor should write with a felt tipped pen or do problems on the chalkboard to make it easier for the learning disabled student to distinguish between factors, numbers and variables. If possible, have tutors ask partially sighted students some of their learning strategies which may help these learning disabled students.

Short-Term Memory/Auditory Processing

Learning disabled students with short-term memory or auditory processing problems, in my opinion, are the second most common group of learning disabled students having problems in mathematics. Students who have a short-term memory problem may not remember the mathematical steps in order, or forget certain problem steps.

These students may remember most of the lecture words but get the words mixed up with other words or have gaps in their memory. These students may not be able to remember facts, understand concepts and write down this information all at the same time. Mathematics notes on problem steps may be in the wrong order or not there at all.

Learning disabled students with auditory processing problems may have a good short-term memory but do not process all the words in the lecture. These students will misinterpret part of the instructor's lecture resulting in gaps in their lecture notes. These same students may also write down misunderstood words that make no sense at all. In either case these students have difficulty understanding the lecture and recording notes.

Since mathematics is considered linear learning, misunderstanding one word or not remembering one word of a concept or formula can cause a misunderstanding of the total concept or formula. The result is that these students are only receiving part of the lecture. This leads to gaps in understanding the instructor and in notes which drastically interfere with learning.

Some of the learning accommodations for learning disabled students with short-term memory or auditory processing difficulties are note-takers, tape recorders, physical proximity and video tapes. As discussed earlier, the note-takers should be from the same mathematics class or at least have the same mathematics teacher. Learning disabled students should be able to understand the note-taker's notes and use these notes to fill in the gaps in their own notes. Learning disabled students should still take some of their own notes; they should not rely totally on the note-taker's notes. Learning disabled students must rework their notes as soon as possible after class to improve their learning. While reworking the notes, the students can write down questions about unclear material to ask the instructor or tutor.

The major problem in using an audio tape recorder is the listening time. Listening to an hour mathematics tape might take an hour and a half. Using an audio tape recorder with a tape counter can help learning disabled students improve their understanding of lecture material with less wasted time. The tape counter indicates how much of the audio tape has passed through the tape player. Students can use the tape counter to indicate the tape number of the beginning and end of a mathematics problem discussion. Later on, that exact discussion can be located on the tape and replayed. This will allow students to fill in the gaps in their notes and better understand the material; listening to the entire tape may frustrate learning disabled students with auditory processing difficulties.

The use of mathematics video tapes as part of the textbook publisher's package, professional video tapes for learning disabled students, or made-by-the-math-department tapes can be helpful. These video tapes can be especially helpful if the tapes have subtitle captions. Students with audio processing problems usually have a visual preferred learning style that is conducive to learning from video tapes.

Video tapes can be rewound and played again as many times as needed to understand the material. Learning disabled students can take notes from the video tape at their own speed. An excellent learning strategy is to watch the mathematics video tape on a difficult mathematics concept right after mathematics class. However, some students may prefer to watch the mathematics video tape right before the mathematics class. This will give the student an overview on the new concepts so he/she can be better prepared to understand the concept during the classroom lecture.

Trained tutors will be required for some of the learning disabled students with auditory processing/short-term memory problems. Tutors must understand that these learning disabled students will only be able to understand part of a lecture. This also applies to the tutorial sessions. These learning disabled students will probably only understand part of the tutor's problem solving explanation at any given time. Tutors should have the learning disabled students repeat back the just-explained facts or concepts. This is one way to make sure learning disabled students understand the facts or concepts.

Fluid Reasoning

Learning disabled students with fluid reasoning (abstract) problems will have extreme difficulty learning mathematics. This could also include students with head injuries. These students usually have poor organizational skills, poor problem solving skills, and trouble understanding causal relationships. The difficulties causing the most problems are poor abstract reasoning and difficulty generalizing from one experience/idea to new situations.

If some of your students have a fluid reasoning learning disability, they may learn how to do a mathematics problem but not be able to generalize that concept to their homework problems. They may demonstrate the knowledge of a mathematics concept one day and forget how to use the same concept the next day. This usually occurs because they are trying to memorize mathematics problem patterns because they cannot understand the mathematics concept.

Learning accommodations for learning disabled students with abstract thinking/reasoning problems are extensive. Most of these students will need a combination of note-takers, tape recorders, handouts, mathematics video tapes, fact sheets, tutors and calculators. Tutors who are trained in helping learning disabled students are the most important learning accommodation. Tutors must start working with these students no later than the second week of class. The tutor should start by giving the student a test covering the previous course material. The tutors can make up questions from the previous chapter reviews and give it to the student. This will give the tutor a better understanding of what the student does not remember from the previous courses. Every tutoring session should start with a review of the material covered in the previous session. After reviewing the previous material, the tutor can start on the new material. Before ending the tutoring session, a review of the new material must be conducted. By following this tutoring procedure, the learning disabled student will retain more but learn less per tutorial session. **Remember: working with students with thinking/reasoning problems will require more patience and tutorial sessions because less is learned in each session.**

Calculators can be another learning accommodation for learning disabled students with fluid reasoning problems. Calculators can reduce some of the problems with the mechanics of mathematics. This will allow the student to have a better focus on mathematics concepts. Many learning disabled students working mathematics problems don't know if they got the problem wrong due to their calculation errors or because they used an incorrect mathematics concept. Calculators can help alleviate this problem so students can focus on applying the correct mathematics concept instead of worrying about the mechanics.

Long-Term Retrieval

Some students have difficulty with mathematics due to a long-term memory retrieval problem. Students may appear to know their basic mathematics skills one day and the next day forget how to do basic calculations. If your students have this type of learning disability, then they may score good grades on quizzes but fail major tests. When they are taught a forgotten concept again, they can relearn it faster than the first time. This means that they retained some of the concept knowledge but not enough to remember the total concept. This long-term memory retrieval problem is different than thinking/reasoning problems. A long-term memory retrieval problem usually applies to the mechanics of doing mathematics problems or not remembering a concept. However, the student understands the logic or reasoning behind the concept. Being told the mathematics concept should allow these students to work that type of problem.

If your students have a long-term memory retrieval problem, then some learning accommodations can be mathematics video tapes, mathematics study skills training, note cards, calculators, handouts, and tutors. These students will have to constantly review difficult mathematics procedures to keep from forgetting them. This can be accomplished by marking certain parts of mathematics video tapes or using note cards for review. Calculators should be used to take the place of the mechanical aspects of mathematics. Tutors can be used to help these students review the parts of mathematics that the student keeps forgetting. Before every mathematics test these students must understand the concept errors of the previous test and know how to use these concepts.

This discussion has focused on learning accommodations for specific types of learning disabilities. Learning disabled students may have problems in one or more of the processing areas. This means a combination of learning accommodations may be needed to enhance their learning. Most learning disabled students have additional learning problems, including poor time management, high test anxiety, social interaction difficulties, and poor test-taking skills. In order to be academically successful, learning disabled students must focus on these additional learning problems as much as their learning accommodations.

Processing deficits or dyscalculia are only two of many learning disabilities that can cause mathematics learning problems. Most learning disabled students who have difficulty in mathematics have visual or auditory processing disorders. These processing disorders can cause an interruption or complete disruption in obtaining lecture material inherent to learning mathematics. Learning disabled students with fluid reasoning problems will have the most difficulty learning mathematics because of their poor ability to grasp mathematics concepts. Learning disabled students with tong-term memory retrieval problems will have difficulty remembering their basic mathematics skills. However, in many cases learning disabled students with the proper accommodations and good mathematics study skills can succeed in learning mathematics.

To make sure your learning disabled students obtain the proper learning accommodations for their disability, refer to Reference A. Reference A lists the appropriate accommodations for different learning disability areas.

Chapter 5

Mathematics Testing Accommodations

5

Appropriate testing accommodations for learning disabled students are used to separate measuring a student's learning disability and measuring the student's mathematics knowledge. This is especially true when time is considered a major factor in measuring mathematics knowledge. In the past, when most learning disabled students were given a mathematics test, the test measured both the degree of their learning disability and their mathematics knowledge. In almost all these cases, the result was a lower mathematics grade which was not a true indicator of the student's mathematics knowledge. This grade frustrated the learning disabled mathematics student and in many situations frustrated the mathematics instructor. The student was frustrated because he/she knew more than the test indicated. The mathematics instructor also became frustrated because the student demonstrated mathematics knowledge in class but failed the test. This section will suggest appropriate testing accommodations for mathematics students with different types of learning disabilities. This is not an all inclusive list but can be considered an appropriate guide.

Visual Processing Speed/Visual Processing

As previously stated, most of the learning disabled students I have helped who had difficulty in mathematics had a visual processing and/or visual speed problem. If this is most of your students' type of learning disability, then timed mathematics tests will cause major problems in determining mathematics knowledge. Let's say your student has this learning disability and starts taking a mathematics test with fifteen problems. Most mathematics students will be able to read each mathematics problem in a matter of seconds. But it will take your student a lot longer. In fact, in might take up to a minute to decipher a complicated mathematics problem and your student still may have misread the mathematics signs. While the other students are working on the first mathematics problem, your student is still making sure he/she read the problem correctly. The extra time needed for reading mathematics problems may cost your student ten minutes.

After reading the mathematics problem, your student starts working on the answer step-by-step. However, it takes your student longer to work the problem due to slower visual processing speed and/or visual processing. Your student is constantly rechecking his/her work for misread symbols or signs. Your student is now on problem eight and has only eight minutes left to complete the test. Your student decides to rush through the remainder of the problems. He/she starts making careless errors and does not finish the last four test questions. Unfortunately your student knew how to work two of the last four problems. Taking more time to work the mathematics problem and checking each step may have cost your student an extra fifteen to twenty minutes. Your student's test is returned with a grade of 58. He/she failed the test. But remember, half of your student's test time was taken up coping with his/her learning disability. **Was this test a true measurement of your student's mathematics knowledge? NO!**

Learning disabled students with visual processing speed or visual processing disorders need to have as much extra time as necessary to compensate for their learning disability. Depending on the type and severity of their learning disability, along with the type of test, these students may need one and a half to two times the normal test time. I have allowed some students up to four times the normal time, if they have a severe visual processing speed and visual processing problem. Don't get in the habit of giving all your learning disabled students the same amount of extended time unless it is until they finish the test.

Another testing accommodation for learning disabled students with visual processing speed or visual processing problems is to enlarge the test questions. By enlarging the test questions and spreading out the questions, learning disabled students are less likely to become confused by similar symbols and numbers. This can also be accomplished by having the mathematics instructor use a felt tipped pen to hand write the test in large numbers and symbols. A test can be enlarged by blowing up the test 150 percent and putting two questions per 14" by 17" sheet of paper. Learning disabled students will have less of a chance of misreading the test question and plenty of room to work the problems.

Learning disabled students should divide the working space of the test page in half. Each problem step should be worked on the left hand side of the page. The calculations that are required for each problem step should be put on the right hand side of the page. Using this system will be less confusing, because students will not mix up problem steps and calculations. Make sure students check every answer of each mathematics problem before turning in their test.

Test readers and/or test proof readers are two more testing accommodations that can help students with visual processing speed or visual processing problems. Test readers are used to make sure these students are correctly reading the test. When students receive their test, they need to read the test out loud to the test reader to compare what is said to the actual problem statement. If the student makes a mistake, then that part of the problem should be written in large print and a different color. If these students do not understand the test question, then the reader can clarify the meaning of the test questions. When clarifying the test question the reader has to make sure not to include information on how to work the problem. As these students finish each problem or after finishing the entire test, the test reader can become a test proof reader. The test proof reader will read each problem, step by step, as the student has written it. If the student disagrees with what the proof reader said, then the student needs to review that problem.

When I was a test proof reader, students usually found at least two major mistakes on test problems, most often in the form of missing sign changes. Many of these students misread signs in the problem steps. If some of your students have a history of reading mathematics problem steps one way then reading them a different way the next time, a test proof reader will be needed. Just using a test reader may present a problem; the test reader needs to read the students' answers back to the student.

Recording the test questions on an audio cassette tape and playing the tape while students read the questions is another testing accommodation. By hearing the test question on tape, your students will be able to make sure they understand the numbers, symbols and words of the mathematics problems. If your students' main problem is just reading the test questions and not misinterpreting their own handwriting, this could be their best testing accommodation.

Learning disabled students may require a private testing area. This testing area may or may not have a chalkboard. Most learning disabled students have problems with distractions while being tested. If your learning disabled students are working with a test reader or test proof reader, then they will need a private testing area so they don't disturb other students.

Some students can think better when walking around and writing problems on the chalkboard. These students may be kinesthetic learners. By writing problems on the chalkboard, there is less chance of misreading signs and working the problem incorrectly. Moreover, these students most likely learned how to work the problem from a chalkboard. If this is true, then recalling how to work the problem may be easier standing at the chalkboard.

Learning disabled students may have difficulty keeping the steps of a mathematics problem in order. You can use regular lined paper or get the paper blown up to 125 percent to make it easier to write on. Another way to use lined paper is to turn it on its side to make the lines into columns. Put a number, sign or variable of the equation in each column. Use the columns to keep them in order. As the problem is worked, the amount of columns will decrease. In most cases, students should end up with three columns containing a variable, equal sign, and number.

These are a few testing accommodations for learning disabled students with visual processing speed/or visual processing problems. This does not mean that all your students will need all these testing accommodations to circumvent their learning disability. The combination of testing accommodations I usually recommend is extended test time, enlarged test and a private testing area. These three testing accommodations have solved most of my learning disabled students' testing problems.

Short-Term Memory/Auditory Processing

Test accommodations for learning disabled students with auditory processing and/or short-term memory problems depend on the type and length of the mathematics test. If the mathematics test contains some type of oral component, then students may need a private testing area with a test reader or tape recorder. If your students have a short-term memory problem, then mathematics tests with story problems, questions with several parts or long written introductions will take more time to complete. Learning disabled students may have to read the test question many times or return to different parts of the question to obtain a full understanding. Meanwhile, the other students are already working on the test problems. These other students will most likely finish before your students and will have time to check their answers. Your students may have taken an additional ten minutes to reread and remember the questions, leaving less time to finish the problems and check their answers. These students will need extended test time.

Learning disabled students with auditory processing difficulties may have similar test taking problems due to their potential reading difficulties. These student need additional time to finish their test in order to measure their mathematics knowledge instead of their reading ability.

Fluid Reasoning/Long-Term Memory Retrieval

Learning disabled students with fluid reasoning problems or long-term memory retrieval problems will need special testing accommodations. If some of your students have this type of learning disability, then test taking time will be a major problem. It will probably take these students longer to remember how to work the problems or to do the mechanics of working the problems. These students will need a calculator to work the problems in order to make sure that their calculations are correct. This will increase the chances of using the correct concept because there is less chance of making careless errors.

Using the *Strategy Cards for Higher Grades* (Nolting, 1988) suggestions and mnemonic and acronym memory cues are also important. In fact, students with fluid reasoning or long-term retrieval problems should be allowed to use a mathematics handout. This handout would contain the important concepts and formulas needed to work the problems. This handout would be similar to the mathematical information allowed for statistics tests. Remember: one of the main purposes for learning mathematics is to be able to use mathematics in the "real world". In the "real world" most mathematics questions would be answered with appropriate resources such as calculators, mathematics textbooks, formularies and a reasonable length of time.

Currently, there is a controversial discussion among mathematicians on the use of calculators in the classroom. The National Council of Teachers of Mathematics has come out with its 1989 guidelines. Part of these guidelines strongly recommends the use of calculators in every mathematics course except arithmetic. There was a recent Office of Civil Rights (Connecticut State Department of Education vs. Office of Civil Rights - 3/13/1980 LOF) case involving the use of calculator by a learning disabled student on a college admission test. The Office of Civil Rights ruled that calculators can be used on admissions tests that do not test basic arithmetic. According to Heyward, Lawton and Associates this case can be generalized to mean that learning disabled students are allowed to use calculators on all mathematics tests except the tests involving arithmetic calculations. In fact, starting in 1992, the SAT test will allow the use of calculators for all students. The use of calculators by learning disabled students having difficulty in mathematics has become just as common as use of the Franklin Spelling computer for English students who have difficulty spelling words when writing an in-class paper.

Potential Test Accommodations Concerns

Many learning disabled students want the least amount of testing accommodations possible. Some of these students may begin the semester by not using testing accommodations. After taking a few mathematics tests they decide to request testing accommo- dations. Some of these students also ask to retake their previous mathematics tests, which is not a reasonable accommodation under Section 504 based on Office of Civil Rights decisions.

Sometimes when students request testing accommodations, the mathematics instructor starts questioning the reason for the accommodations. For example, I had a student wait until the final exam to request a private testing area and extended testing time. She requested the testing accommodation in plenty of time for accommodation to be arranged. The next day I received a call from the mathematics instructor questioning why the student was now requesting testing accommodations. The instructor explained that she never knew the student was learning disabled and was doing "B" to "A" work in the course. It appeared to the teacher that the student was not having difficulty finishing the mathematics tests and it might not be fair to the other students to give her extended time. I then explained to the mathematics instructor that our philosophy is to give students the least amount of testing accommodations necessary, but to make sure we are not testing the students' learning disability instead of their mathematics knowledge.

In every case it is the learning disabled student's decision to request testing accommodations. I continued to explain to the instructor that this student was allowed testing accommodations. The student had decided to wait until the final exam to request extended test time for two major reasons. First, the final exam was going to count as two major test scores or about 40% of her grade and would make the difference between an "A" or "B". The second reason was that the final exam was going to last two hours instead of the regular test time of about 50 minutes. This student had been compensating for her visual process/speed disorder by working fast and catching most of her mistakes. However, taking a departmental final exam which is more difficult than her regular test created an anxiety state. The test anxiety is related to her fear of not being able to complete all the problems and check them within the time limit. As you may know, test anxiety can increase the effects of a student's learning disability.

I then told the mathematics instructor that this student could have been using testing accommodations throughout the semester but chose not to. The mathematics instructor agreed with me and now understood more about learning and testing accommodations.

Some learning disabled students may have to be encouraged to use their testing accommodations-especially students making above average grades, who are shy, or are fearful of their instructors. Sometimes these students would rather risk getting a lower grade than be treated "special". On the other hand, other learning disabled students want every testing accommodation allowed and some that are not allowed! Both types of students need to be counseled about their learning disability and their appropriate testing accommodations.

Learning disabled students need to be taught about the effect of their learning disability on testing. They need to be able to explain to the instructor how testing accommodations can circumvent their learning disability and measure their true mathematics knowledge. Most mathematics instructors will try to understand the student's learning disability and arrange the testing accommodations. This arrangement may be having the student take the test at the Office of Disabled Students.

Unfortunately not all mathematics instructors are knowledgeable about learning disabilities and understand the reasons for testing accommodations. Some mathematics instructors provide testing accommodations by having the student start the test in the classroom and finish it somewhere else. This procedure may appear to be equitable but it can lead to improper test accommodations. For example, mathematics instructors have had students follow them back to their office to finish the test in their office or the department office. This is not an appropriate testing accommodation due to on-going distractions and the possibility of an unsaid implication to hurry up and finish the test. Many colleges do not allow this as a testing accommodation any more.

In other situations the mathematics instructors may try to persuade disabled students not to request the testing accommodations. Another problem occurs when the mathematics instructor refuses to allow testing accommodations. These instructors may say that they do not believe in learning disabilities, that it is their academic freedom not to allow testing accommodations, or that they do not allow that type of "babying" in their class.

You need to make your students aware of these potential testing accommodation problems. If these problems do occur, the Office of Disabled Students must be informed. It is the prevailing professional opinion that the counselors in the Office of Disabled Students are the most qualified to make the decisions on appropriate testing accommodations--not the mathematics instructor. **Remember: appropriate testing accommodations are a learning disabled student's civil right and it is against the law to deny appropriate testing accommodations.** To select the appropriate testing accommodation for your learning disabled students, refer to Reference B.

Alternative Test Forms

In most cases, the instructor will select the test form that will be the most effective way to evaluate a student's mathematics skills. The test formats are usually based on tradition and ease of scoring. Unfortunately some of the test forms will cause difficulty for certain learning disabled students. Even though large print tests and audio cassette tape test forms have been discussed, there are other test forms that could be used. Some of these other test forms are: oral tests, video monitor enlarged tests, computer enlarged tests, computer speech enhanced tests, and take-home tests. Each test form has its advantages and disadvantages for the student and instructor. You need to discuss these alternate test forms with your disabled students to discover which form is likely to give the best indication of the students' mathematics knowledge rather than their learning disability. For an overview of alternative math test forms turn to Reference B.

Sometimes you may not know the effects of your students' learning disabilities and what testing accommodations they should receive. In this situation I would suggest the following testing accommodations: private testing area, extended time, and enlarged test. You then must find out the types of learning disabilities of your students to make sure that they are receiving appropriate testing accommodations.

There are different types of testing accommodations for different types of learning disabilities. Some learning disabled students may not require any testing accommodations while other learning disabled students will require extensive testing accommodations. Make sure to discuss appropriate test-taking accommodations with each of your disabled students. Help your students practice explaining the reasons for their test-taking accommodations. This practice will prepare your students to explain the necessity for their test-taking accommodations so that the instructor will have a better understanding of the effects of their learning disability.

Chapter 6

Student Case Studies

6

This next section is about learning disabled students whose mathematics grades have been improved. In my opinion, most learning disabled students need a combination of mathematics study skills, learning accommodations, and testing accommodations in order to reach their mathematics learning potential. Some learning disabled students only needed the Math Study Skills course, using the *Winning At Math* (Nolting, 1988) textbook and the audio cassette tape, *How To Reduce Test Anxiety* (Nolting, 1987), to become successful in mathematics. **Remember: math study skills are especially important because your students are already having difficulty receiving/processing lecture/textbook information.** Mathematics study skills learning techniques can compensate for some of your students' learning problems. Having the appropriate accommodations will not do much good if your students do not have good mathematics studying and test-taking skills. On the other hand, without the appropriate learning and testing accommodations your students cannot demonstrate their mathematics knowledge.

The following case studies are an example of combining mathematics study skills, learning accommodations, and testing accommodations to help students become successful in mathematics. I will proceed from the least difficult to the most difficult case histories. These are only a few case histories of learning disabled students whom I have assisted with mathematics learning.

Student A

Student A is a thirty-year-old male who was in my Math Study Skills course and was having difficulty passing his basic mathematics tests. It appeared from his class participation and homework assignments that he was using the correct mathematics study skills and test-taking skills. While talking to him, he indicated that he had a D average on his mathematics tests six weeks into the semester. His main problem was not having enough time to finish his mathematics tests. He said that it took a lot of time to write down each problem step and he constantly checked for careless mistakes. Doing his mathematics homework took him longer than most students, but it appeared that he understood the mathematics concepts. He also said that in high school he had reading problems and was in some type of special education class. Some of his family members also had reading problems. Reviewing his mathematics test revealed that the completed problems were mostly correct. The problems toward the end of the test had careless errors or were not complete.

Student A had a 97 intelligence score which is average. His intelligence rank was in the forty-third percentile. The *WJ-R Test of Cognitive Ability* indicated his visual processing speed test score was extremely low. The visual processing speed standard score was 64 which is at the fifth grade level. These scores translate into a first percentile visual processing speed rank score. This means that 99 percent of the other students had better visual processing speed skills than Student A. A fiftieth percentile score is a middle score meaning that half of the other scores are below and half are above that score. There was over a thirty standard score point difference between his visual processing speed score and his IQ.

His auditory processing standard score was a 64 which is at the second grade level representing a second percentile rank. This means that 98 percent of the students were auditorially processing the mathematics lectures better than Student A. This also means that his intelligence score may be higher than indicated because part of his IQ test was verbal. This low test score definitely indicates that Student A was having difficulty understanding the mathematics lecture. His two other processing scores were average: visual processing (which is separate from visual processing speed) standard score was 109 (seventy third percentile), and short-term memory standard score was 97 (forty third percentile).

The test results indicated Student A had difficulty working quickly under pressure without making careless mistakes. This meant that he took more time on mathematics tests and made more careless errors. Student A also had difficulty understanding some of the lecture information. On the other hand, Student A had good skills in recognizing figures, finding incomplete figures and understanding spacial configurations. He could quickly recognize complicated mathematics equations and formulas. He also had an average short-term memory standard score.

As you may remember, a learning disability in Florida is:

> A disorder in one or more of the basic psychological or neurological processes involved in understanding or in using the spoken or written language. Disorders may be manifested in listening, thinking, reading, writing, spelling, or performing arithmetic calculations.

Student A has a learning disability because his visual processing speed disorder and auditory processing disorder dramatically affect his ability to process information and take tests. Student A's visual processing speed disorder decreased his writing speed and accuracy of written material. This problem manifested itself in his not performing fast arithmetic calculations during tests and constant rechecking of problem steps. It also blocked his ability to understand mathematics problems written on the board because his concentration was on accurate note-taking instead of the instructor's explanations on working problem steps.

Student A's auditory processing disorder affects his ability to understand the mathematics lecture material (listening) and oral test directions. Tutorial services for Student A will also be affected because the tutor might have to explain the reasons for doing mathematics problem steps several times. The auditory processing disorder mainly affects Student A's ability to learn mathematics through lectures.

The accommodations offered to Student A were note-takers, private testing area, and extended test time. Student A elected to use only a private testing area and extended test time. Student A made a "B" on his next mathematics test and a "B" on his final mathematics test for a "C" average mathematics course grade. He is now in the next level mathematics course and at mid-term is still making good grades and not using a note-taker. I suggested to Student A that he would eventually need a note-taker due to the increased difficulty of learning mathematics.

Student A's case is a good example of how the combination of mathematics study skills and testing accommodations resulted in a student becoming successful in mathematics instead of failing the course. This was the first time in Student A's life that he felt successful in a mathematics course.

Student B

Student B is a middle-aged woman who indicated problems with concentration, distractions, test anxiety, reversing letter/numbers, and short term auditory/visual memory. She has a life-long history of mathematics learning problems. She also said that many of her family members had similar learning problems. With all these problems, she had a 4.0 grade point average. She was taking her first mathematics class (basic algebra) in about twenty years. She was in the basic mathematics class for a while and decided to drop the course due to anxiety and learning problems. She was referred to the learning center to improve her basic mathematics skills. She continued to take the Math Study Skills course and was referred to the Assessment Center for a learning styles survey.

An intelligence test score indicated she had an IQ of 98 which is average. The scores from the *WJ-R Test of Cognitive Ability* indicated a severe visual processing speed learning problem and a possible auditory processing learning problem. Even though she had average intelligence, her visual processing speed standard score was 55 which is fourth grade level or a one-tenth percentile rank score. This means that over 99 percent of the students had better visual processing speed than she did. Her auditory processing standard score was 86 which is at the fifth grade level or a seventeenth percentile rank score. This means that 83 percent of college students had better auditory processing skills than she did. Her short-term memory and fluid reasoning standard scores were low average. However, her visual processing standard score was 105 which is at the grade thirteen level or a sixty-second percentile rank score. Her mathematics aptitude standard score was 86 and her broad math (mathematics achievement) standard score was 93. There was not a significant difference between her mathematics achievement score and her mathematics aptitude or intelligence score.

Student B was learning disabled because her visual processing speed disorder prevented her from accurately writing material as fast as other students. This problem manifested itself in not performing arithmetic calculations as quickly as other students during tests. This caused her problems on timed mathematics tests. It also blocked her from understanding mathematics problems written on the board. She took more time to copy the material off the board leaving less time to understand the mathematics problem.

Her low Auditory Processing score suggested a low ability for understanding sound patterns (words) under distracting conditions. She also had difficulty in putting sounds together to make words. Student B had problems understanding lectures in a noisy classroom and confused one word for another. She had difficulty taking notes. She left out words in her notes or put the wrong words in her notes, causing problems in understanding mathematics concepts. Her low-average fluid reasoning ability cluster score caused some abstract learning problems.

During the remaining part of the semester Student B continued in the Math Study Skills course and attended the learning center. She made an "A" in the Math Study Skills course. She relearned her arithmetic skills and algebra skills.

The next semester Student B again enrolled into the basic algebra course. The accommodations for Student B were a note-taker for her mathematics course, extended test time, and a private quiet testing area. Student B made an "A" on her first mathematics test which was the first "A" she had ever made on a mathematics test. She made the highest final exam score in the course and an "A" in the course. Combining previous mathematics learning, mathematics study skills, appropriate testing and learning accommodations with persistence made her a successful mathematics student.

Student C

Student C is dyslexic and extended test time was suggested as an accommodation. Student C performed poorly on her first major mathematics test and had a history of failing mathematics classes. After a second interview with Student C, additional testing was conducted to reveal all aspects of her learning disability. Since she was not in the Math Study Skills course, additional affective surveys were given.

Student C had test anxiety at the ninety-ninth percentile level, excellent general study skills and study attitude, external locus of control, and an auditory numeric learning style. To decrease the test anxiety she was instructed to practice with the audio cassette tape *How To Reduce Test Anxiety* (Nolting, 1988). She was also given instructions on the best procedures for auditory learning.

Student C had average intelligence (90 IQ) but her intelligence score was probably depressed by her processing disorders. Student C had an extreme disorder in visual processing speed, short-term memory, and visual processing. Her processing speed standard score was 64, which is at the fifth grade level, or a first percentile rank score. Her short-term memory standard score was 72 which is at the second grade level or a third percentile rank score. Her visual processing was a 74 standard score at the third grade level, which is the fourth percentile. Her auditory processing standard score was 80 indicating a fifth grade level or a tenth percentile rank.

Student C had a learning disability in visual processing speed, short-term memory and visual processing. The effects of visual processing speed on learning and test taking have already been discussed. Student C's short-term memory deficit caused problems in storing mathematics information long enough to use it. For example: when a mathematics instructor explained five steps to solve a mathematics equation, she had difficulty remembering all the steps long enough to write them down. Her visual processing disorder was also a concern because of her difficulty in recognizing the meaning of a combination of mathematics symbols and numbers. Student C had major problems remembering, understanding, and writing down mathematics concepts all at the same time.

The accommodations for Student C were expanded to include a note-taker, tutor and test reader/proof reader. The note-taker allowed her to concentrate more on the mathematics instructor's explanation while still taking some notes. The mathematics tutor explained the mathematics concepts on a individual basis to enhance Student C's learning. The test reader made sure Student C was correctly reading the test questions and her test answers. With these accommodations Student C made an "A" in her mathematics course. The next semester Student C took the Math Study Skills course and her last required mathematics course. Student C made a "B" in her last mathematics course and has graduated. She has been accepted at the University of South Florida and will major in psychology. Her success came with extreme persistence, math study skills training, learning accommodations and testing accommodations.

Student D

Student D is a woman in her mid thirties who had been trying to complete her mathematics requirement to graduate for ten years. She passed the required arithmetic course on the third attempt. While attempting to pass the basic algebra course at the local adult high school and college for the fifth time, Student D was referred to the Assessment Center. Student D had a 3.0 grade point average with 59 completed hours toward her Associate of Arts degree. She was majoring in English. The only courses she needed to graduated were two more mathematics courses and six hours of electives.

Student D has average intelligence based on the *Slosson Intelligence Test* (IQ = 99) and a 95 Full Scale intelligence score from the *WAIS-R* test. The *WAIS-R* verbal score was 91 and the *WAIS-R* performance score was 100. The only significant subtest weakness was picture completion, which measures sequencing in a logical order.

Student D's *WJ-R* test scores were compared to the *WAIS-R* full scale intelligence score of 95. Student D's short-term memory standard score was 92 which is at the eighth grade level or a twenty-ninth percentile rank score. Her visual processing speed standard score was 89 which is at the ninth grade level or a twenty-fourth percentile rank score. The auditory processing standard score was 89 indicating a six grade level and a twenty-third percentile rank score. The visual processing standard score was 102, suggesting a thirteenth grade level or a fifty-sixth percentile rank score. Even though most of Student D's standard scores were in the low average range her main learning problem was in fluid reasoning. Her fluid reasoning standard score was 69, which is at the third grade level or a second percentile rank score. Her mathematics aptitude standard score was a 79 or eighth percentile which also indicated a mathematics learning problem. This score means that 92 percent of college students have a better mathematics learning potential than she does. However, student D's broad math achievement standard score was 102 which is at the fourteenth grade level or a fifty-seventh percentile rank score. Student D's actual mathematics achievement score was 1.5 standard deviations above her expected mathematics achievement score.

Student D has a learning disability in fluid reasoning (abstract reasoning) or thinking (Florida definition) due to a 26 point difference between fluid reasoning score and her IQ score. This fluid reasoning deficit was further verified by the low score in the *WAIS-R* subtest (picture completion) which measures logical sequence ability.

This assessment information was explained to Student D and she was offered learning and testing accommodations. It was suggested that Student D tape record the mathematics class, use a tutor, watch the video tapes for learning disabled students, make up note cards, and do her homework problems in different colors for tracking purposes. She was also taught how to improve her mathematics study skills. The testing accommodations were extended time, use of a calculator, and a private testing area.

Student D's main comment about the interpretation of the results was that after ten years of frustration and anger she was glad that someone finally told her what her problem was. With these learning and testing accommodations given to Student D, and her studying about forty hours a week, she made a "B" in the mathematics course. However, at that time I knew she would probably not pass the next mathematics course and a course substitution would have to be requested for her last mathematics course.

As mentioned earlier, a learning disability based on a fluid reasoning deficit is difficult to overcome in comparison to processing disorders. Processing disorders can be compensated for through learning aids but it is very difficult to improve someone's innate reasoning ability.

Student D is now in the next algebra course, which is the mathematics course she needs to graduate. We have already begun the process for the mathematics course substitution based on the college guidelines set by the Florida Legislature. Student D has now decided to formally request a course substitution. The course substitution will probably be for a Philosophy of Logic course or a general computer course.

Two questions came into mind after reviewing Student D's case. The first question was, "With such a low fluid reasoning score and a low mathematics achievement score how did she manage to have a broad math achievement score of 102?" In other words, how did she learn enough mathematics to be at the 14th grade level, which is one and a half standard deviations above her expected mathematics achievement. Part of the question has already been answered by knowing that she studied mathematics 40 hours a week. In addition, she had mathematics tutors who constantly reviewed her mathematics skills and taught her new skills. She also took the Math Study Skills course to maximize her learning/test taking skills and had individual learning strategy sessions.

The second question was, "What kept this student continuing to take ten mathematics courses, one after another with little success?" Just like many learning disabled students, Student D is very persistent and will go to any length to improve her education. Student D's academic achievement motivation and internal locus of control drive her to be successful at every learning task. The personal support from her family also gave her a reason to continue taking mathematics courses. Her will to succeed is stronger than that of most learning disabled students and should be directed in a positive area.

The summation of all of Student D's effort helped her pass the first basic algebra course. However, the real question is, "How much more mathematics can she learn?" It is my opinion that Student D will not be able to go much higher than one and a half standard deviations above her predicted mathematics achievement. In other words, she has learned about as much mathematics as she will ever learn. **Remember: even with the best learning/testing accommodations some students will need to substitute mathematics courses in order to meet their graduation requirement.**

Student Anti-X

The next student case presented the most challenge in helping a learning disabled student learn mathematics. At the suggestion of Bill Reineke, a mathematics lab director, I am calling this student Anti-X instead of Student E. Bill said that Anti-X was an appropriate name for this student because of his tremendous dislike of mathematics. Anti represents what a person dislikes and X represents mathematics.

In 1987, Anti-X was referred to me because of his mathematics learning problems. He had passed the basic arithmetic course but was failing the basic algebra course. His major was computer science and he was making "A"s and "B"s in all his computer programming courses. He indicated several times that he did not like mathematics. He wanted to know why he needed mathematics, since he was making good grades in his computer courses. After talking to him about mathematics learning skills, I requested he take the next Math Study Skills course. Since learning mathematics was similar to learning computer programming, I thought mathematics study skills and test taking skills were his main problem. Anti-X withdrew from the basic algebra course and took the second part of the semester-long Math Study Skills course. He made an "A" in the Math Study Skills course and the following semester repeated the basic algebra course. He did not pass the mathematics course.

The following semester he re-enrolled in the algebra course and began to have learning difficulties. Anti-X returned to my office saying that he was studying mathematics about twenty hours a week but was failing the course. However, he was still making "A"s and "B"s in his computer courses. I decided that there must be something else affecting his ability to learn mathematics.

Anti-X completed a questionnaire for a learning disability evaluation. During the interview he indicated some motor nerve impairment which slowed down his handwriting. He also indicated problems with visual perception speed and memory problems. He indicated that the main problem was putting mathematics concepts into long-term memory. As a side comment, he said that he became disoriented in new places, until he learned his way around. His grade point average was 3.22.

Anti-X completed the *Survey of Study Habits and Attitudes* (Brown and Holtzman, 1984), *Math Study Skills Evaluation* (Nolting, 1989), old *Woodcock-Johnson Psycho-Educational Battery* (WJPETB) (1977), and the *Wechsler Adult Intelligence Scale-R* (1981). Anti-X had a score at the ninety-fifth and ninety-ninth percentile respectively on Study Habits and Study Attitudes which indicated excellent study skills and attitudes. He made a score of 82 percent on the Math Study Skills Evaluation which indicated good mathematics study skills. There was a twenty-three point difference between his Verbal IQ (86) and Performance IQ (63). His Verbal IQ was the highest and was used as the best indicator of his ability. His verbal IQ was in the low average range. His IQ score was probably depressed due to his slight motor impairment and processing disorders.

The old *WJPETB* combined the Visual Speed and Visual Perception test. Anti-X's visual speed/perception standard score was 67 which was at the fourth grade level, which is at the first percentile. Anti-X's Reasoning standard score was a 71, which is at the second grade level score, or a third percentile rank score. The reasoning score measures nonverbal abstract reasoning and problem solving skills.

The memory score of the old *WJPETB* measures short-term auditory memory. Anti-X's memory standard score was a 105 indicating a twelfth (12.9) grade level score which is at the sixty-fourth percentile. His verbal ability standard score was 100 which is at the fiftieth percentile. His mathematics aptitude standard score was 85 which is at the sixteenth percentile and his mathematics achievement standard score was 92 which is the thirtieth percentile. There was no significant difference between Anti-X's mathematics aptitude standard score and his mathematics achievement standard score. Based on this data, Anti-X should not have passed his first mathematics course or be making "A"s and "B"s in his computer programming courses.

Due to his documented motor nerve impairment, Anti-X was classified as having a physical disability. However, the testing data indicated that he had a learning problem due to the severe difference between his IQ score and his visual speed/perception cluster score. If Anti-X had not been classified as being physically disabled, he would have qualified under Florida law as a learning disabled student.

Since there was very little applied research in the area of mathematics learning accommodations, I depended on my imagination to suggest accommodations for Anti-X. Learning accommodations changed several times based on different mathematics concepts and mathematics courses. Anti-X did not use all of these learning accommodations at the same time but selected the learning accommodations that were best suited for learning the mathematics concept.

Anti-X was having difficulties keeping his mathematics homework in some legible organization. He would bring his homework problems for the tutors to review, but the tutors couldn't understand what he wrote. Anti-X improved his homework technique by drawing a line down the center of the page. On the left side of the page, he wrote the problem steps, and on the right side of the page he did the mathematical calculations for each step. This kept the problem steps separate from the calculations.

Anti-X misread some of the mathematical notations and signs while doing his homework. I had Anti-X use a felt tipped pen which made the mathematics numbers and symbols larger and easier to read. Anti-X used this system for a while but abandoned it for another system. Anti-X now used a ball point pen that had four colors: red, black, green, and blue. When doing mathematics homework he used different colors for different variables, numbers, factors, exponents, negative signs or positive signs. This does not mean that he had a different color for each part of the mathematics problem. He used blue for the parts of the mathematics problem he already knew how to work and used the other pen colors for the new parts. Now he could follow each mathematics operation by pen color, until the problem was completed. Even though it took Anti-X longer to work the problems, this way decreased his visual processing errors and made it easier to work the problems.

Anti-X worked with several mathematics tutors as an additional learning aid. The best tutor Anti-X had was a very intelligent visually impaired tutor named Mike. Mike had already developed techniques for learning and taking mathematics tests that circumvented his visual disability. Mike shared these techniques with Anti-X. For example, we discovered that Anti-X's writing speed was not capable of keeping up with the speed at which he mentally processed mathematics problems. When Anti-X's mental processing got ahead of his writing, he made careless errors and lost track of the problem solving procedures. Anti-X was asked to solve this problem in two ways. One way to reduce his mental speed was to write each mathematics problem step in a different colored pen. This reduced Anti-X's mental problem solving speed, resulting in fewer problem steps omitted. This procedure was then combined with having Anti-X write, in English, what he was doing beside each mathematics step. This procedure not only slowed down Anti-X's mental processing speed but also helped him learn the reasons for each problem step.

Teaching Anti-X how to do mathematics graphing problems was another challenge. Mike determined the best method for Anti-X to learn graphing was using the "shift method". Rather than merely plotting points and drawing a line though them, Mike gave Anti-X a concept model about parabolas. Anti-X was taught how to draw a simple parabola on a graph. Using the standard parabola equation of $y = (x - h)^2$, Anti-X was taught that the sign inside the parenthesis determines if the parabola is to the left or right of the Y axis. A negative sign shifts the parabola to the right and a positive sign shifts the equation to the left. To better understand the graphing of a parabola, Anti-X drew a basic parabola. He then changed pen colors and drew a new parabola with a shift to the right. He changed pen colors again and drew a another parabola with shift to the left. The different colored parabolas made it easier to understand graphing parabolas. By looking at the sign of h, Anti-X could tell which side of the y axis to put the graph on. Finally these graphs were put on index cards with different pen colors to be understood and memorized.

Anti-X was also having difficulty solving problems involving complex numbers such as $(3a^4 b^2 c^{-2})^{-4} (2a^{-2} bc^3)^2$. He would look at the total problem and become confused and not know where to start. To solve this problem a rectangular piece of cardboard was cut out to lay over the first part of the problem. This allowed Anti-X to isolate different factors or signs and concentrate on one part of the problem at a time. This eliminated most of the visual confusion caused by his learning disability.

Anti-X had difficulty understanding and working word problems. He was taught to break down the word problem into two parts. He wrote the word phrases for each part of the problem and put the algebraic expressions under each word phrase. For example: if the word phrase was "a number decreased by seven", then he wrote "x - 7" under it. Anti-X used the Strategy Card Number 12, *Translating English Words Into Algebraic Expressions* (Nolting, 1988) and learned how to write the algebraic expression for word problems.

Many different learning accommodations were tried with Anti-X to make the mathematics material accessible to him. Not all these learning accommodations were used at the same time. Some others were tried but did not work. The team approach by Anti-X, his tutor and myself successfully developed learning accommodations that helped Anti-X improve his ability to learn mathematics.

Anti-X's testing accommodations were changed several times. Basically they included extended test time, private testing area, and test proof readers. Some of his testing accommodations were similar to his learning accommodations. Since Anti-X had difficulty with his handwriting due to a motor nerve disability and a visual speed/

perception learning disability, he was given three times the normal testing time. The private testing area was in a comfortable room with a chalkboard. He was allowed to take breaks under supervised conditions. This allowed him more control over his anxiety. Sometimes test proof readers were used to read the test and his answers to him. Later on, test proof readers were discontinued.

Using an alternate test form was a major breakthrough for Anti-X. At first, we used the regular teacher-made test with a reader, but the test format was so visually crowded that he had difficulty understanding the problems and writing the steps to the problems. On several occasions Anti-X correctly solved test problems on scratch paper and then miscopied them back to his test paper. I then had the test blown up 150 percent so that four problems were on a page. This helped Anti-X but did not totally solve the visual processing problem. Anti-X's test was then blown up another 150 percent with one or two problems per 14" by 17" page. This test format eliminated the misread problem errors and eliminated the need for a test proof reader.

However, another problem occurred when Anti-X mixed up his steps in solving mathematics equations with the calculations used to solve the equations. For example, after working three-fourths of a mathematics problem, Anti-X could not find the last step used to solve the equation. His calculations were so mixed up he could not find where he left off working the equation. To solve this problem I had Anti-X divide his test page in half by folding it down the middle. The left side of the page was used to work the problem, step by step. The right side of the page was to do the calculations. This test taking strategy solved the problem of mixing up the problem steps and calculations.

When Anti-X began to have tests on graphing equations, his visual process/speed disability became a major problem. Anti-X could not tell the difference between the points on the X and Y axes. To solve this problem, regular graph paper was blown up 150 percent with one sheet of graph paper per problem. On the graph paper Anti-X first drew the X axis in red, then drew the Y axis in blue. Each point on the X axis was indicated by a red mark. Each point on the Y axis was indicated by a blue mark. This marking system made it possible to locate the coordinates on the graph. After having several tests with graphing problems, Anti-X only needed to draw in the X and Y axis with different colors to mark the coordinates.

Anti-X next took a statistics course. His statistics course was causing him some visual problems, especially with bar graphs. To solve that problem, Anti-X made each bar graph and the measurements along each bar graph a different color. He used graph paper that was enlarged 150 percent for his test. Plotting points on the bar graph did take him extra time but he got most of the graphs correct.

Anti-X had difficulty using some of the complicated statistics formulas. He could understand the concept behind the formulas but had a difficult time putting in the correct number for the variable. For example, Anti-X had trouble using the Pearson r correlation formula. To solve this problem I had Anti-X make each variable a different color. He made the number that was supposed to replace the variable the same color as the variable. For example the variable N was written in red and represents the number of subjects in the correlation. The number 20 which represented the variable N was also written in red. Now he could follow the color coded numbers and understand how they relate to the formula. To better understand statistics formulas, Anti-X used different colors to distinguish different parts of the formulas.

Understanding how to use the bell curve proved to be another challenge. To understand that the area under the bell curve represents the probability of an event to occur was conveyed though a tactical map. A piece of graph paper that consisted of equal square boxes was enlarged 150 percent. A bell curve was drawn on the graph paper with a line bisecting the center of the curve. One large rubber band was placed along the edges of the bell curve. Anti-X was told that there were equal numbers of squares on each side of the bisected curve. He was also told to move his fingers along the rubber band to get a feeling of how the area under the curve changed as he reached the top of the curve. This made it easier for him to understand the future concept of hypothesis testing.

When Anti-X was conducting *z-tests* and *t-tests* he was instructed to write the null hypothesis statement in one color and the rejection of the null hypothesis statement in a different color. He then colored in the critical z or t areas of the curve the same color as the rejection null hypothesis statement. This made it easier for him to recognize the critical *t-score* area or *z-core* areas. He then placed the calculated *z-score* or *t-score* in the correct place on the bottom of the curve. If the *z-score* or *t-score* was in the colored area he was told to reject the null hypothesis. If the *z-score* or *t-score* was not in the colored area he was told to accept the null hypothesis. To further help Anti-X find the critical values he was given separate enlarged *t* distribution tables for one and two tailed tests. The results of this work was that Anti-X made a "B" in his statistics course.

Currently Anti-X has finished all of his required algebra classes and the statistics course. He has a computer job waiting for him upon graduation. This success was a joint effort of good instruction from the mathematics department and learning/ testing accommodations from the Assessment Center. The two instructors most responsible for teaching Anti-X mathematics and statistics were Mr. Bill Savage (Successful Math Study Skills author) and the Chair of the mathematics department. Anti-X has become a successful mathematics student through persistence, good instruction, and accommodations.

In fact, Anti-X has become so successful that Bill Reineke asked him to be a mathematics tutor. Anti-X now likes mathematics and has become a mathematics tutor. I have dropped Anti from his fictitious name. He can now be called Student X.

Student X has helped other learning disabled students who have learning disabilities similar to his. He has completed some individual tutoring in statistics but knows his visual processing limits.

Some people might question how Student X learned college algebra with his reasoning at a second grade level. In reviewing the four subtests that make up the reasoning cluster, I saw two of those tests require a high level of visual processing. These two reasoning subtest scores were below the two other subtest scores. In other words, the reasoning test was also measuring his poor visual perception/speed, instead of just measuring his reasoning ability. This lowered his reasoning score and was not a true indication of his reasoning ability. This one incident does not mean that all the reasoning scores are biased by students with poor visual processing, however.

Chapter 7

Referral Indicators, Information References, Legal Ramifications and Course Substitutions

7

Referral Indicators

A major part of helping students with mathematics learning problems is appropriate referrals from mathematics instructors. Mathematics instructors need to be trained how to recognize signs of learning disabilities. There are several signs that might indicate a learning disability exists. One sign is the student who demonstrates an understanding of mathematics concepts in the classroom but fails the mathematics test. Another sign is the student whose homework or test problems look like "chicken scratch" with no reasonable order. A third sign is the student who makes "A"s on his/her quizzes but makes low "D"s or "F"s on major tests. A fourth sign is the student who spends hours on his/her homework and in the mathematics lab but still makes "D"s or "F"s.

One of the best signs of a learning disability was explained by a mathematics lab director. He brought over one of his intermediate algebra students and said that her test results indicated only a few points were missed on the first three-fourths of the test. However, the student missed almost all the points on the last part of the test even though she appeared to know the material. He stated that it appeared she just ran out of time. This is a good indication of a possible visual processing speed disorder. The student's test results indicated a 115 IQ with visual processing speed standard score of 80 which is at the seventh grade level or a tenth percentile rank. With extended time this student improved her next mathematics test score. These learning disability signs also could indicate students who were placed in the wrong mathematics course or have some type of physical disability.

Even if some of the referrals made prove to be incorrect, you will help those who need it, and not hurt those who don't. With training, mathematics instructors can do a good job of identifying possible learning disabled students to be referred to the Disabled Student Coordinator. A team approach to helping learning disabled students become more successful in mathematics is the best approach.

Information References

Training mathematics instructors and administrators is a necessary requirement for a better understanding of learning disabilities. Training college personnel on learning disabilities is a difficult task that can be made easier by using the material distributed by the Association on Handicapped Student Service Programs in Postsecondary Education (AHSSPPE). AHSSPPE is an organization from which to obtain information, training programs, workshops, publications and resources on learning disabled college students. The mission of AHSSPPE is to provide leadership, focus, and expertise for professionals working with disabled students. Membership in this organization is a must for any counselor, instructor or administrator working with disabled students. I have been a member of AHSSPPE for years and have found its yearly conferences excellent, with diverse workshops and excellent net-working opportunities.

Reference C has a list of AHSSPPE publications and a membership application. For workshop training I would recommend the *AHSSPPE Inservice Education Kit* (1988) and the *Survival Kit for Learning Disabled Students in Higher Education* (1988). To train your tutors I would suggest *Assisting College Students with Learning Disabilities: A Tutor's Manual* (1990/second printing). *Testing Accommodations for Students with Disabilities* (1990) provides excellent guidelines for establishing or fine-tuning a creditable, viable adaptive testing program. If you are interested in expanding your knowledge on learning disabilities and other types of disabilities I would suggest *AHSSPPE's Annotated Bibliography of Information Resources* (1991/second edition).

When conducting workshops on disabilities, I would suggest inviting Dr. Jane Jarrow, Executive Director of AHSSPPE, as the guest speaker. Dr. Jarrow can set a positive tone on helping learning disabled students become more successful in college. Dr. Jarrow will also give you insight on how to conduct future workshops and how to work with the administrators on providing reasonable accommodations to learning disabled students.

Legal Ramifications

Now that you have a better understanding of the effects of different learning disabilities on mathematics and appropriate learning/testing accommodations, the legal responsibilities of the college need to be discussed. During the last several years I have read my share of court cases regarding disabled students and still lack a good understanding of their implications. Last year I started using *Access To 504* (1990) by Heyward, Lawton, and Associates, Ltd. (Reference D). *Access To 504* is a computer program that is the most comprehensive treatment of Section 504 and its implementing regulations presently available. The authors have used their 22 years of experience in enforcement of Section 504 through the Office of Civil Rights to develop an effective resource tool. *Access To 504* also has a written training manual and reference guide that can supplement the computer program.

I now depend on *Access To 504* laws to understand the legal responsibilities for colleges and universities. I have called Ms. Heyward and Mr. Lawton several times to obtain clarification on different legal aspects of Section 504. They have always been helpful and had given me useful information. *Access To 504* is an excellent resource to convince administrators of the legal responsibilities of the college or university.

There are several common questions regarding Section 504 that I have been asked while conducting training workshops. Some of these questions are very simple to answer while other questions require interpretation of court cases. Some of these questions are:

1. Is not having money to provide note takers for learning disabled students a legal excuse for not providing this service?
2. Is it permissible to give out a list of all the learning disabled students to the mathematics instructors so they can identify them in their classrooms?
3. Can a mathematics instructor tell a learning disabled student that he/she cannot tape record the lecture?
4. Can mathematics instructors decide how much extended test time a learning disabled student should receive?
5. Can instructors themselves be held legally responsible for violating a student's civil rights?

To answer these question I contacted Ms. Heyward who is an expert attorney on Section 504. Below are Ms. Heyward's answers to question one through five:

1. Institutions are ultimately responsible for providing note-taking services to learning disabled students regardless of the financial status unless another service agency provides that service or the institution can claim undue hardship. However, no institution to my knowledge has successfully argued that note-taking is an undue hardship.

2. No. The regulations state that it is the learning disabled students' decision to identify themselves to the instructor. The disabled student counselor cannot tell instructors which students have disabilities unless permission is given by the student.

3. No. The regulations specifically list the use of tape recorders as a academic adjustment. Disabled students are allowed to use tape recorders unless the instructor can find an alternative way to accommodate that student.

4. The mathematics instructor alone should not make the decision if a disabled student should receive extra time or the amount of extra time given. The decision on the amount of extended time needs to be made with the mathematics instructor, student, and the disabled student counselor. The decision should be based on the type and extent of the learning disability and the type of test.

5. If you are talking just about the Office Of Civil Rights then the institution is held legally responsible for the violations, not the instructor. However, the University of California at Berkeley case set a precedent that instructors can be sued in Civil Court for discriminating against learning disabled students.

Ms. Heyward answered these questions based her knowledge as a civil rights attorney. However, the answers to three of these question were already explained in *Access To 504.* Now you may have a better understanding about the usefulness of *Access To 504.*

The legal ramifications of Section 504 are changing with each court case. Institutions are becoming more aware of their legal responsibilities to disabled students. To keep up with these changes I would recommend subscribing to the *Disability Accommodation Digest* publication from AHSSPPE. This publication is being written by Heyward, Lawton, and Associates and will have the latest rulings and information on the legal aspects of Section 504 and learning disabled students. *Disability Accommodation Digest* also has an "Ask the Expert" column where counselors can mail in their own disability related questions.

Course Substitutions

Altering classroom instruction is another way to help learning disabled students improve their mathematics learning. Research suggests that altering teaching styles to accommodate learning disabled students can also improve the learning of non-disabled students. However, in some cases learning disabled students, even with all the suggested accommodations, may not pass the required mathematics courses. These students, who have advanced as far as they can in mathematics, can be allowed course substitutions. In Florida, the Department of Education has mandated that universities and colleges set up course substitution guidelines for learning disabled students. Some of the course substitutions for mathematics are business mathematics, computer sciences, general sciences, logic, and accounting. Computer science can be a course which teaches students how to run computer programs such as word processing programs. The logic course is usually taught by the philosophy department and is the basis of modern mathematics.

The course selected for substitution should have the least effect on the student's learning disability and be closely related to the student's major or concepts representing mathematics. Consult the Office of Disable Students for more information about the rules on course substitutions.

Summary

A learning disability is a neurological disorder that cannot be cured. Learning disabled students must be given accommodations to make sure they have appropriate access to learning materials. Learning disabled students also need testing accommodations to measure their achievement so it is not affected by their learning disability. This right is guaranteed by Section 504 of the Rehabilitation Act of 1973. Section 504 is a federal civil rights law that supersedes any state, institution, college or university policy that discriminates against disabled students.

Helping learning disabled students become successful in mathematics is a group responsibility for administrators, counselors, and mathematics instructors. Disabled student counselors need to work with administrators and mathematics faculty to help them understand the effects of a learning disability. The counselors can explain to the learning disabled students the effects of their disabilities. Then appropriate learning accommodations and testing accommodations can be given to the learning disabled students. This group effort will help learning disabled students reach their mathematics potential and graduate from college.

REFERENCES

Reference A

Suggested Mathematics Learning Accommodations For Specific Learning Disabilities

These are suggestions for mathematics learning accommodations based on specific learning disabilities. These learning accommodations have been proven successful with many learning disabled students. Not all of these learning accommodations are usually needed for most of the learning disabled students. Even using all these learning accommodations may not circumvent every student's learning disability. However, in every case the learning disabled students must be taught math study skills to help compensate for their learning problems.

Learning Accommodations

Visual Processing Speed/Visual Processing

Note-takers • Re-work notes • Tape recorder with tape counter • Large print handouts • Large print copies of important textbook pages • Taped textbook • Turn note book sideways • Take notes in different pen colors • Trained tutors

Short-Term Memory/Auditory Processing

Note-takers • Re-work notes • Tape recorder with tape counter • Physical proximity • Math video tapes • Trained tutor • Tape record important tutor explanations

Fluid Reasoning\Long-Term Retrieval

Note-takers • Re-work notes • Tape recorder with tape counter • Handouts • Math video tapes • Facts sheets (flash cards) • Color coded problem steps • Trained tutor • Tape record important tutor explanations • Strategy Cards for Higher Grades • Calculators

---

Reference B

Suggested Mathematics Testing Accommodations For Specific Learning Disabilities

These are suggestions for mathematics testing accommodations based on a specific learning disability. These accommodations have been proven effective with learning disabled students when trying to measure their mathematics knowledge instead of their learning disability. In most cases every listed testing accommodation will not be needed for most learning disabled students. However, depending on the severity of the learning disability, even all these testing accommodations may not totally circumvent the student's learning disability.

Testing Accommodations

Visual Processing Speed/Visual Processing

Extended time • Private test area • Quiet test area • Enlarged test questions • Test readers • Test proof readers • Recording test on audio tape • Kinesthetic options (chalk board) • Specially lined/oriented paper • Color coded math equations • Calculators

Short-Term Memory/Audio Processing

Extended time • Private test area • Quiet test area • Recording test on audio tape

Fluid Reasoning/Long-Term Memory

Extended time • Private test area • Quiet test area • Fact sheet handouts • Strategy Cards for Higher Grades • Clarification of test questions • Calculators

Alternative Test Formats

Oral test • Enlarge test on a video monitor • Computer enlarged test • Computer speech enhanced test • Take home test

---

Reference C

Publications and Products Catalog

Association on Handicapped Student Service
Programs in Postsecondary Education
P.O. Box 21192
Columbus, OH 43221-0192
614-488-4972

The Association of Handicapped Student Service Programs in Postsecondary Education is a multinational, not-for-profit organization committed to promoting full participation of individuals with disabilities in postsecondary education. The Association's numerous training programs, workshops, publications, and conferences are planned and developed by its elected officials and governing board and carried out by the full-time Executive Director and staff.

PUBLICATIONS

Unlocking the Doors: Making the Transition to Postsecondary Education (1987)

A publication designed to assist high school students with learning disabilities, their parents, and teachers in preparing for the transition to the postsecondary setting. $12.50 (Member price...$7.50)

Programming for College Students with Learning Disabilities (1990/Second printing)

Developed by Anna Gajar, Pennsylvania State University, this publication offers a model for institutions that are beginning support services components. $31.00 (Member price... $21.00).

Assisting College Students with Learning Disabilities: A Tutor's Manual (1990/Second printing)

This manual is designed for use by service providers and tutors working with students with learning disabilities. Includes sample tutoring program for spelling and looseleaf sheets for reproduction purposes. $26.00 (Member price...$16.50)

AHSSPPE's Annotated Bibliography of Information Resources (1991/Second edition)

Published in looseleaf form or on Macintosh Hypercard 2.0, this invaluable resource provides descriptions and contact information on more than 250 books, journals, association, media productions, and products. $39.00 (Member price...$27.00) for looseleaf manual OR Hypercard version. $54.50 (Member price $42.00) for BOTH looseleaf manual AND Hypercard versions.

Survival Kit for Learning Disabled Students in Higher Education (1988)

Original, camera-ready copy is provided for reproduction of seven pamphlets designed to orient the student with a learning disability. $32.95 (Member price...$27.95)

AHSSPPE Inservice Education Kit (1988)

All the handouts and documentation necessary to conduct inservice training for the postsecondary community regarding the inclusion of students with disabilities in campus life. $32.95 (Member price...$27.95)

Testing Accommodations for Students with Disabilities (1990)

This guide is written for service providers who are responsible for arranging test accommodations for students with disabilities in higher education. $23.00 (Member price...$15.50)

Membership Categories

Active Professional $75.00
Affiliate $65.00
Institutional $187.50
Student $35.00

For more information about AHSSPPE membership and publications, contact:

614-488-4972 (V/TDD)
Fax: 614-488-1174

Reference D

ACCESS TO 504

**Heyward, Lawton, and Associates
PO BOX 1436, Atlanta, GA 30301
(404) 636-8485 - (800) 525-9220**

Heyward, Lawton and Associates presents ACCESS TO 504, the most comprehensive reference and research tool available to address the many questions facing professionals today regarding the rights of handicapped persons and the responsibilities of educational institutions and agencies.

Through court case listings, as well as key word and citation indexing, ACCESS TO 504 explains the basic principles that apply to each provision of the regulation. It also outlines specific situations where those principles have been developed, and applied by the courts and the U.S. Department of Education, Office for Civil Rights (OCR).

ACCESS To 504 will assist you in:

- maintaining 504 compliance,
- answering day-to-day questions about 504 with confidence
- minimizing litigation, and
- minimizing the intervention of Federal enforcement agencies

This comprehensive guide to handicap/disability law is available in two formats: a computer software program or a hard copy manual. Also, purchasers will receive supplemental updates, automatically, which will provide instant access to the most current information available.

Hard Copy Manual

You may purchase the complete reference guide ($150) which addresses all subparts of the regulations, or individual subparts*, such as:

- **A Postsecondary Education Guide ($95)**
- **The Elementary and secondary Education Guide ($95)**
- **The Employment Guide ($95)**

Sample pages of the manual are available upon request.

*all individual subpart guides include discussions of the general provisions and program accessability subparts.

Computer Software Program

The program permits you to access information by using four distinct indices:

- **Case Summaries**
- **Key Words**
- **Regulatory Citations**
- **Individual Subparts**

Price $350

A demonstration disk which gives you a preview of the software is available upon request.

ADDITIONAL SERVICES

Equity Audits

Heyward, Lawton and Associates will conduct in-depth reviews of the pertinent policies and procedures of districts/institutions having difficulty complying with Section 504. Compliance problems will be identified and strategies will be developed to address them.

Complaints Assessment and Resolution

Heyward, Lawton and Associates staff will review complaints of disability/handicap discrimination and provide advice regarding the merits of the complaints and recommend possible solutions. We will also assist institutions in preparing for Federal investigations.

Information Service

Our staff will answer your questions regarding disability discrimination law. Our database and resources are available to you for on-the-spot answers to your compliance questions. We will also perform specialized research upon request.

Disability Accommodation Digest

A quarterly newsletter which contains information and articles of interest to persons concerned with the implementation and enforcement of Section 504 and the Rehabilitation Act of 1973 and the Americans with Disabilities Act of 1990.

BIBLIOGRAPHY

Atkinson, R., & Shiffrin, R. (1968). Human memory: A proposed system and its control processes. In K. W. Spence & J. T. Spence (Eds.). **The Psychology of Learning and Motivation** (Vol. 2). New York: Academic Press.

Badian, N. A. (1988). The prediction of good and poor reading before kindergarten entry: A nine-year follow up. **Journal of Learning Disabilities,** 21, 98-103, 123.

Batchelor, E. S., Gray, J. W., & Dean, R. S. (1990). Empirical testing of a cognitive model to account for neuropsychological functioning underlying arithmetic problem solving. **Journal of Learning Disabilities,** 23, 38-42.

Bauer, R. (1982). Information processing as a way of understanding and diagnosing learning disabilities. **Topics in Learning & Learning Disabilities,** 2(2), 46-53.

Bley, N. S., & Thornton, C. A. (1981). **Teaching Mathematics to the Learning Disabled.** Rockville, MD: Aspen System.

Blalock, J. W. (1980). Persistent auditory language deficits adults with learning disabilities. **Journal of Learning Disabilities,** 15, 604-609.

Bourne, L., & Ekstrand, B. (1985). **Psychology : Its Principles and Meanings,** New York, NY: Holt, Rinehart, and Winston.

Brown, W. F. & Holtzman, S. H. (1984). SSHA manual: Survey of study habits and attitudes. New York: The Psychology Corporation.

Cawley, J. F. (1984). An integrative approach to needs of learning-disabled children: Expanded use of mathematics. In J. F. Cawley (Ed.), **Developmental Teaching of Mathematics for the Learning Disabled** (pp. 81-94). Rockville, MD: Aspen System.

Cox, L.S. (1975). Systematic errors in the four vertical algorithms in normal and handicapped populations. **Journal for Research in Mathematics Education,** 6, 202-220.

Daniels, P.R. (1983). **Teaching the Gifted/Learning Disabled Child.** Austin, TX: PRO-ED.

Dinnel, D., Glover, J., & Ronning R. (1984). A provisional mathematical problem solving model. **Bulletin of the Psychonomics Society,** 22, 459-462.

Engelmann, S., Carnine, D., & Steely, D. (1991). Making connections in mathematics. **Journal of Learning Disabilities,** 24,292-303.

Enright, B. E. (1987). Basic mathematics. In J.S. Choate, T.Z. Bennett, B. E. Enright, L. J. Miller, J. A. Poteet, & T. A. Ranks (Eds.), **Assessing and Programming Basic Curriculum Skills** (pp. 121-145). Boston: Ally and Bacon.

Fleischner, J., & Garnette, K. (1980). **Arithmetic Learning Disabilities: A Literature Review.** Research review Series (1979-1980). Research Institution for the Study of Learning Disabilities, Columbia University, NY. (ERIC Document Reproduction Service No. ED 210 843).

Hammill, D. H. (1990). On defining learning disabilities: An emerging consensus. **Journal of Learning Disabilities,** 23, 74-83.

Hessler, G. (1984). **Use and Interpretation of the Woodcock-Johnson Psycho-Educational Battery,** Allen, TX: DLM Teaching Resources.

Jarrow, J. (1987). Integration of individuals with disabilities in higher education: A review of the literature, **Journal of Postsecondary Education & Disability,** 5(2) 38-57.

Kirby, J.R. & Becker, L. D. (1988). Cognitive components of learning problems in arithmetic. **Remedial and Special Education,** 9(5), 7-16.

Kochnower, J., Richardson, E., & DiBenedetto, B. (1983). A comparison of the phonic decoding ability of normal and learning disabled children. **Journal of Learning Disabilities,** 16, 348-351.

Malcolm, C. B., Polatajko, H.J., & Simmon, J. (1990). A descriptive study of adults with suspected learning disabilities. **Journal of Learning Disabilities,** 23, 518-520.

Mather, N. (1991). **An Instructional Guide to the Woodcock-Johnson Psycho-Educational Battery.** Bradon, VT: Clinical Psychology Publishing Co., Inc.

Maters, L. F., & Mori, A. A. (1986). **Teaching Secondary Students With Mild Learning and Behavior Problems: Methods, Materials, Strategies.** Rockville, MD: Aspen Systems.

Meyers, M. J. (1987, November). LD students: Clarifications and recommendations. **Middle School Journal,** 27-30.

Miller, J. H., & Milam, C. P. (1987). Multiplication and division errors committed by learning disabled students. **Learning Disabilities Research,** 2, 119-122.

Miller R. J., Snider, B. & Rzonca, C. (1990). Variables related to the decision of young adults with learning disabilities to participate in postsecondary education. **Journal of Learning Disabilities,** 23, 349-354.

Morris, M., & Leuenberger, J. (1990). A report of cognitive, academic, and linguistic profiles for college students with and without learning disabilities. **Journal of Learning Disabilities,** 23, 355-360.

Nolting, P. D. (1987). **How to Reduce Test Anxiety.** Pompano Beach, FL: Academic Success Press.

Nolting, P. D. (1988). **Winning at Math: Your Guide to Learning Math the Quick and Easy Way.** Pompano Beach, FL: Academic Success Press.

Nolting, P. D. (1989). **Strategy Cards for Higher Grades.** Pompano Beach, FL: Academic Success Press.

Nolting, P. D. (1989). **Math Study Skills Evaluation Computer Software.** Pompano Beach, FL: Academic Success Press.

Nolting, P. D. (1991). **Winning at Math: Your Guide to Learning Mathematics Through Effective Study Skills.** Pompano Beach, FL: Academic Success Press.

Rourke, B. P., & Strang, J. D. (1983). Subtypes of reading and arithmetic disabilities: A neuropsychological analysis. In M.J. Rutter (Ed.), **Developmental Neuropsychiatry,** (pp 473-488). New York: Guiford Press.

Salvia, J., Gajar, A., Gajria, M., & Salvia, S. (1988). A comparison of WAIS-R profiles of not disabled college freshmen and college students with learning disabilities. **Journal of Learning Disabilities,** 21, 632-636.

Strang, J.D., & Rourke, B. P. (1983). Concepts formation/non-verbal reasoning abilities of children who exhibit specific academic problems with arithmetic. **Journal of Clinical Child Psychology,** 12, 33-39.

Strang, J. D., & Rourke, B. P. (1985). Arithmetic disability subtypes: The neuropsychological significance of specific learning impairments in childhood. In B.K. Rourke (Ed.), **Essentials of Subtypes Analysis** (pp. 167-183). New York: Guiford Press.

Thornton, C. A. & Bley, N. S. (1982). Problem solving: Help in the right direction for learning disabled student. **Arithmetic Teacher,** 29(6), 26-27, 38-41.

Thornton, C. A., & Toohey, M. A. (1985). Basic math facts: Guidelines for teaching and learning. **Learning Disabilities Focus,** 1(1), 44-57.

Waldron, K.A., & Saphire, D. G. (1989). Perceptual and academic patterns of learning-disabled/gifted students. Manuscript submitted for publication.

Waldron, K. A., & Saphire, D. G. (1990). An analysis of WISC-R factors for gifted students with learning disabilities. **Journal of Learning Disabilities,** 23, 491-498.

Wanger, R. K. (1986). Phonological processing abilities and reading: Implications for disabled readers. **Journal of Learning Disabilities,** 19, 623-630.

Wechsler, D. (1981). **Wechsler Adult Intelligence Scale-Revised.** New York, NY: The Psychological Corporation.

White, W., Alley, G., Deshler, D. Schumaker, J., Warner, M., & Clark, F. (1982). Are there learning disabilities after high school? **Exceptional Children,** 49 273-274.

Woodcock, W. (1977). **Woodcock-Johnson psycho-educational battery achievement tests.** Allen, TX: DLM Teaching Resources.

Woodcock, W. , & Johnson, M. (1989). **Woodcock-Johnson Psycho-educational Battery Achievement Tests - Revised.** Allen, TX: DLM Teaching Resources.

AUTHOR BIOGRAPHICAL DATA

Dr. Paul Nolting

Over the past fifteen years Learning Specialist Dr. Paul Nolting has helped thousands of students improve their mathematics learning and obtain better grades. Dr. Nolting is an expert in assessing mathematics learning problems - from study skills to learning disabilities - and developing effective learning strategies and testing accommodations.

He is a national consultant and trainer of math study skills and of learning and testing accommodations for learning disabled students. He has conducted national training grant workshops on learning disabilities for the Association on Handicapped Student Service Programs in Postsecondary Education and for the University of Wyoming 1991 Training Grant of Basic Skills. Dr. Nolting has conducted numerous national conference workshops on learning disabilities and mathematics for the National Developmental Education Association and the National Council of Educational Opportunity Association.

Dr. Nolting holds an M.S. degree in Counseling and Human Systems from Florida State University and a Ph.D. degree in Counselor Education from the University of South Florida. His book, *Winning at Math: Your Guide to Learning Mathematics Through Successful Study Skills* was selected Book of the Year by the National Association of Independent Publishers. "The strength of the book is the way the writer leads a reluctant student through a course from choosing a teacher to preparing for the final examination, "says Mathematics Teacher, a publication of the National Council of Teachers of Mathematics.

His two audio cassettes, *How to Reduce Test Anxiety* and *How to Ace Tests* were also winners of awards in the National Association of Independent Publishers competition. "Dr. Nolting," says Publisher's Report, "is an innovative and outstanding educator and learning specialist."

A key speaker at numerous regional and national education conferences and conventions, Dr. Nolting has been widely acclaimed for his abilities to communicate with faculty and students on the subject of improving mathematics learning.